Many Variations of Mahler Measures

The Mahler measure is a fascinating notion and an exciting topic in contemporary mathematics, interconnecting with subjects as diverse as number theory, analysis, arithmetic geometry, special functions and random walks. This friendly and concise introduction to the Mahler measure is a valuable resource for both graduate courses and self-study. It provides the reader with the necessary background material, before presenting recent achievements and remaining challenges in the field.

The first part introduces the univariate Mahler measure and addresses Lehmer's problem, and then discusses techniques of reducing multivariate measures to hypergeometric functions. The second part touches on the novelties of the subject, especially the relation with elliptic curves, modular forms and special values of L-functions. Finally, the appendix presents the modern definition of motivic cohomology and regulator maps, as well as Deligne–Beilinson cohomology. The text includes many exercises to test comprehension and to challenge readers of all abilities.

T0201369

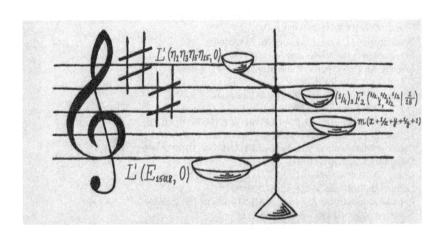

Australian Mathematical Society Lecture Series: 28

Many Variations of Mahler Measures
A Lasting Symphony

FRANÇOIS BRUNAULT
École Normale Supérieure de Lyon

WADIM ZUDILIN
Radboud Universiteit Nijmegen

CAMBRIDGE
UNIVERSITY PRESS

CAMBRIDGE
UNIVERSITY PRESS

University Printing House, Cambridge CB2 8BS, United Kingdom

One Liberty Plaza, 20th Floor, New York, NY 10006, USA

477 Williamstown Road, Port Melbourne, VIC 3207, Australia

314–321, 3rd Floor, Plot 3, Splendor Forum, Jasola District Centre, New Delhi – 110025, India

79 Anson Road, #06–04/06, Singapore 079906

Cambridge University Press is part of the University of Cambridge.

It furthers the University's mission by disseminating knowledge in the pursuit of education, learning, and research at the highest international levels of excellence.

www.cambridge.org
Information on this title: www.cambridge.org/9781108794459
DOI:10.1017/9781108885553

First published 2020

Printed in the United Kingdom by TJ International Ltd, Padstow Cornwall

A catalogue record for this publication is available from the British Library.

ISBN 978-1-108-79445-9 Paperback

*À mon père, qui a œuvré pour que
chaque personne se réalise pleinement*

To Our Long-awaited Great Achievements

Contents

Preface

Kurt Mahler, Fellow of the Royal Society, is an illustrious part of Australia's mathematical heritage. His route to prominence, handicapped by tuberculosis in childhood, entailed completion of a PhD thesis under the great Carl Ludwig Siegel, shortly followed by his escape from Nazi Germany to England, and finally acceptance of a professorship at the newly opened Institute of Advanced Studies of the Australian National University in Canberra. At the beginning of the 1960s, at the very time of his move to Australia, Mahler [129–131] became interested in a particular height function of uni- and multivariate polynomials — a *measure*, its analytical representations and applications in number theory. In [130] a new version and a new proof of Gelfond's inequality relating the height of a product to the product of heights was given. This inequality is at the heart of one of the most successful methods in transcendental number theory, which led to the resolution of the seventh Hilbert problem independently by Gelfond and Schneider, and laid the groundwork for Alan Baker's method of linear forms in logarithms developed from the mid-1960s — Baker was awarded the Fields Medal in 1970 for this work. Years later, the original scope of the Mahler measure was expanded significantly after the discovery of its deep links to algebraic geometry and K-theory, in particular to Beilinson's conjectures. These interrelations generated a body of challenging problems, some recently resolved but many remaining open.

The set of notes and exercises below grew from numerous talks and lectures by the authors on Mahler measures, special values of L-functions, regulators and modular forms. This material was crystallised during the masterclass on the topic at the University of Copenhagen, Denmark, during the last week of August 2018. The set-up of two courses there was based on what are now Chapters 1–6 (lectured by W.Z.) and Chapters 7–10 (lectured by F.B.) of this book, with the lectures alternated. We take this opportunity to thank heartily

Fabien Pazuki and Riccardo Pengo, the organisers of this conference, as well as the participants for the inspiring and lively discussions around the lectures.

The book can still be viewed as consisting of two parts. The first part (Chapters 1–6) mainly touches the essential basics for the univariate Mahler measure and things related to Lehmer's problem (also known as Lehmer's question), then accelerates to multivariate settings and discusses techniques of reducing the measure to hypergeometric functions. Some L-functions (or L-values) related to (hyper)elliptic curves and modular forms show up in the background along the way. Then the discussion smoothly transitions into the second part of the book (Chapters 7–10), where the details of recent novelties of multivariable Mahler measures are given. Finally, the appendix presents the modern definition of motivic cohomology and regulator maps, as well as the example of Deligne–Beilinson cohomology.

The text is supplemented with many exercises of different complexity, and the chapters include additional ones that can help the reader to extend their horizon of understanding of the topic and its links to other mathematical gems. All this makes the book a unique and comprehensive introduction to a developing area, which has numerous links with practically any other part of mathematics and, at the same time, suffers from lack of systematic exposition. Apart from many references on the Mahler measure spread out in the literature (and cited in the bibliography list with care), we can point out the books [83, 176] and reviews [93, 198, 199] where certain aspects of the topic are covered.

This text is a friendly and concise introduction to the subject of Mahler measure, a potential source for a graduate course or self-study. There are definitely other aspects of Mahler measure to be discussed, but we have decided to stay realistic rather than encyclopaedic while going through a new — highly specialised! — subject. One reason behind the involvement of Mahler measure in so many areas of mathematics is that it bears the burden of a height function, the reason we reserve as an excuse for not covering all those numerous (and quite important) applications and links. The latter, for example, include applications to algebraic dynamics (well treated in [83]) and remarkable links to Szegő's limit theorems and integrable systems of statistical mechanics [178, 211].

During the writing process we have received valuable support and feedback from our friends, colleagues and collaborators, whom we would like to thank: Marie José Bertin, Denis-Charles Cisinski, Frédéric Déglise, Antonin Guilloux, Matilde Lalín, Hang Liu, Boaz Moerman, Michael Neururer, Fabien Pazuki, Riccardo Pengo, Berend Ringeling, Jörg Wildeshaus. Special thanks go to Michael Neururer for providing us with a sage code for the Riemann surface pictures, to Ralf Hemmecke and Silviu Radu for supplying us with compact expressions for f_{21} and f_{45} in Chapter 6, as well as to Jörg Wildeshaus for allowing us to use the notes from his course at the summer school on special values of L-functions, organised at ÉNS Lyon in June 2014. We offer our heartfelt thanks to Roger Astley, Clare Dennison and Anna Scriven at Cambridge University Press, who provided expert support to bring the manuscript to publication. We are indebted to the copy-editor Alison Durham whose constructive feedback and professional work positively impacted the appearance of the book. The picture on the cover of this book was created by Victor Zudilin, whom we thank warmly for this creative and inspiring illustration of the topic of the book.

We hope that an allusion to the great composer Gustav Mahler in the subtitle of the book does not cause any confusion. Though the great mathematician Kurt Mahler seems to be not related to his famous namesake, his mathematics creature — the Mahler measure — is an artistic work comparable to a symphony in music. Enjoy its sounds!

François Brunault and Wadim Zudilin
Lyon, Nijmegen and Newcastle

1

Some basics

We first agree on customary terminology and notation for polynomials and algebraic numbers.

A polynomial $P(x) = a_d x^d + a_{d-1} x^{d-1} + \cdots + a_1 x + a_0$ of degree d is said to be monic if its leading coefficient is $a_d = 1$. If all coefficients of $P(x)$ are integral, $P(x) \in \mathbb{Z}[x]$, and do not possess a common integral factor, $P(x)$ is said to be primitive. If $P(x) \in \mathbb{Z}[x]$ is not primitive or can be represented as the product of two polynomials from $\mathbb{Z}[x]$ of degrees less than $d = \deg P$, the polynomial $P(x)$ is called reducible; otherwise, it is irreducible (over $\mathbb{Z}$). Besides the degree, standard characteristics of polynomials are its length $L(P) = |a_d| + |a_{d-1}| + \cdots + |a_0|$ and its height $H(P) = \max_{0 \le j \le d} |a_j|$.

A (real or complex) number α is algebraic if it happens to be a zero of a non-zero polynomial with integral coefficients. To each algebraic number α one can assign its minimal polynomial $P(x)$ — a (primitive!) irreducible $P(x)$ such that $P(\alpha) = 0$; the polynomial is well defined up to sign. The degree, length and (naive) height of α are then defined as the degree, length and height, respectively, of its minimal polynomial $P(x)$. Furthermore, all zeros $\alpha_1 = \alpha$, $\alpha_2, \ldots, \alpha_d$ of the minimal polynomial $P(x)$ are called (algebraic) conjugates of α, so that $P(x) = a_d \prod_{j=1}^{d} (x - \alpha_j)$; in the case that they are all real, α is said to be a totally real algebraic number. If the minimal polynomial of α is monic, α is said to be an algebraic integer.

1.1 Kronecker's theorem

Note that if $Q(\alpha) = 0$ for *some* (not necessarily minimal) monic polynomial $Q(x)$ with integral coefficients, then α is an algebraic integer; this is an immediate consequence of the Gauss lemma. Examples of algebraic integers are given by roots of unity $\zeta_n^k = e^{2\pi i k/n}$ for $k = 1, \ldots, n$; they are zeros of the monic polynomial $x^n - 1$.

Exercise 1.1 Show that, for two algebraic integers α and β, their sum $\alpha + \beta$ and product $\alpha\beta$ are algebraic integers as well.

Proposition 1.1 (Kronecker's theorem) *If $\alpha_1 = \alpha$ is a non-zero algebraic integer with all its conjugates α_j inside the unit disc, $|\alpha_j| \le 1$ for $j = 1, \ldots, d$, then α is a root of unity.*

Proof   Let $P(x) = \prod_{j=1}^{d}(x - \alpha_j) \in \mathbb{Z}[x]$ denote the minimal polynomial of α; it is monic, because α is an algebraic integer. For each $n = 1, 2, \ldots$, consider the polynomial $P_n(x) = \prod_{j=1}^{d}(x - \alpha_j^n)$, whose roots are the nth powers of α_j, $j = 1, \ldots, d$. The coefficients of the polynomial $P_n(x^n) = \prod_{j=1}^{d} \prod_{k=1}^{n}(x - \alpha_j \zeta_n^k)$ are integral, hence so are the coefficients of $P_n(x)$ itself. On the other hand, the coefficients are absolutely bounded from above in view of $|\alpha_j^n| \le |\alpha_j| \le 1$ for $j = 1, \ldots, d$. This means that the set of polynomials $\{P_n(x) : n = 1, 2, \ldots\}$ is finite; in other words, $P_n(x) = P_m(x)$ and

$$\{\alpha_1^n, \alpha_2^n, \ldots, \alpha_d^n\} = \{\alpha_1^m, \alpha_2^m, \ldots, \alpha_d^m\} \tag{1.1}$$

for some positive $n \ne m$. Changing the order of elements in the multi-sets if necessary, we can conclude that for some ℓ, $1 \le \ell \le d$, we have

$$\alpha_1^n = \alpha_2^m, \ \alpha_2^n = \alpha_3^m, \ \ldots, \ \alpha_{\ell-1}^n = \alpha_\ell^m, \ \alpha_\ell^n = \alpha_1^m;$$

eliminating $\alpha_2, \ldots, \alpha_\ell$ when $\ell > 1$, we arrive at $\alpha_1^{n^\ell} = \alpha_1^{m^\ell}$, so that $\alpha = \alpha_1$ is an $(m^\ell - n^\ell)$th root of unity.   □

Exercise 1.2 If $\alpha^n = \beta^m$ for an algebraic number $\alpha \ne 0$ and its conjugate β, where integers n, m are such that $|n| \ne |m|$, then α is a root of unity.

Hint   If $\alpha_1, \ldots, \alpha_d$ are algebraic conjugates of α (and β), the Galois group of $\mathbb{Q}(\alpha_1, \ldots, \alpha_d)$ over $\mathbb{Q}$ extends the single equality $\alpha^n = \beta^m$ to the equality of multi-sets (1.1).   □

A polynomial is said to be cyclotomic if all its roots are roots of unity. The minimal polynomial of an nth root of unity $\zeta_n^k = e^{2\pi i k/n}$, where $\gcd(k, n) = 1$, is given by the nth cyclotomic polynomial

$$\Phi_n(x) = \prod_{\substack{k=1 \\ \gcd(k,n)=1}}^{n} (x - \zeta_n^k)$$

of degree $\varphi(n)$, Euler's totient function.

Exercise 1.3 Show that

$$x^n - 1 = \prod_{d|n} \Phi_d(x),$$

where the product is taken over all positive divisors d of n. In particular, the polynomial

$$\frac{x^n - 1}{x - 1} = x^{n-1} + x^{n-2} + \cdots + x + 1, \quad n = 2, 3, \ldots$$

is irreducible if and only if n is prime.

For a polynomial $P(x)$ of degree d, write $P^*(x) = x^d P(1/x)$. If $P(x) = P^*(x)$, we say that $P(x)$ is reciprocal.

Exercise 1.4 Show that a cyclotomic polynomial $P(x)$ satisfies $P^*(x) = \pm P(x)$. If, in addition, $P(1) \neq 0$, then $P(x)$ is reciprocal.

1.2 Factorisation of cyclotomic expressions

It is a classical fact (cf. also Exercise 1.3 with $x = 2$) that all primes of the form $2^n - 1$, $n = 2, 3, \ldots$, correspond to prime $n = p$. On the other hand, the primality of n does not guarantee $2^n - 1$ to be prime, as the following example demonstrates: $2^{11} - 1 = 23 \cdot 89$. *Primes* of the form $M_p = 2^p - 1$ for p prime are known as Mersenne primes; it remains open whether there are infinitely or finitely many such primes. There are 51 Mersenne primes now known (as of February 2019), with the largest one $M_{82,589,933}$ found in December 2018 [138]; it also happens to be the largest prime known. The latter fact is not a coincidence: primes originating from the values of cyclotomic polynomials possess special primality testing.

Exercise 1.5 If p is an odd prime, then any prime q that divides $2^p - 1$ must be congruent to ± 1 mod 8.

Exercise 1.6 (Lucas–Lehmer test) Define the sequence $s_0 = 4$, s_1, s_2, $\ldots$ recursively by setting $s_n = s_{n-1}^2 - 2$ for $n \geq 1$. For p an odd prime, show that the number $M_p = 2^p - 1$ is prime if and only if $s_{p-2} \equiv 0$ mod M_p.

Based on the earlier work of Pierce [152], in 1933 Derrick Henry Lehmer [122] developed primality testing of numbers of the form

$$\Delta_n(P) = \prod_{j=1}^{d} (\alpha_j^n - 1),$$

where $P(x) = \prod_{j=1}^{d} (x - \alpha_j)$ is a monic polynomial with integral coefficients.

In particular, he demonstrated that

$$\Delta_{113}(x^3 - x - 1) = 63{,}088{,}004{,}325{,}217$$

and

$$\Delta_{127}(x^3 - x - 1) = 3{,}233{,}514{,}251{,}032{,}733$$

are primes. Note that in the proof of Proposition 1.1, $\Delta_n(P)$ were cast as $\pm P_n(1)$.

Theorem 1.2 ([122]) *For $P \in \mathbb{Z}[x]$ monic, $P(0)P(1) \neq 0$, the growth of integers $\Delta_n(P)$ as $n \to \infty$ is given by*

$$M(P) = \limsup_{n \to \infty} |\Delta_n(P)|^{1/n} = \prod_{j=1}^{d} \max\{1, |\alpha_j|\},$$

where $\alpha_1, \ldots, \alpha_d$ are the zeros of $P(x)$.

Furthermore, if the sequence $\Delta_n(P)$, $n = 1, 2, \ldots$ is not periodic, then the absolute value of $\Delta_n(P)$ unboundedly increases with n.

Proof The first part follows from

$$\limsup_{n \to \infty} |\alpha^n - 1|^{1/n} = \max\{1, |\alpha|\}$$

for $\alpha \neq 1$.

If the sequence $|\Delta_n(P)|$ (of integers), $n = 1, 2, \ldots$, is bounded, then the limit superior is 1, hence $|\alpha_j| \leq 1$ from the first part of the theorem. By Proposition 1.1 the polynomial $P(x)$ is cyclotomic, hence $|\Delta_n(P)| = |P_n(1)|$ is periodic. $\qquad \square$

Exercise 1.7 Show that if $P(x)$ is a reciprocal polynomial, then $|\Delta_n(P)|/|\Delta_1(P)|$ is the square of an integer for all odd n, and $|\Delta_n(P)/\Delta_2(P)|$ is the square of an integer for all even n.

Hint (Lalín) Use $\alpha^n - 1 = \alpha^{n/2}(\alpha^{n/2} - \alpha^{-n/2})$ for n even and

$$\frac{\alpha^n - 1}{\alpha - 1} = \alpha^{(n-1)/2} \sum_{j=-(n-1)/2}^{(n-1)/2} \alpha^j$$

for n odd, and the fact that if α is a zero of a reciprocal polynomial $P(x)$ then so is $1/\alpha$. $\qquad \square$

Lehmer further suggests [122] some heuristics that in order to obtain large primes from the factorisation of $\Delta_n(P)$, it is advantageous to have the increase of the sequence very slow, in other words, to have $M(P)$ in Theorem 1.2 as small as possible.

Lehmer's problem If ε is a positive quantity, find a monic polynomial $P(x)$ with integer coefficients such that the absolute value of the product of those roots of P which lie outside the unit circle, lies between 1 and $1 + \varepsilon$.

Lehmer gives an example of a small such value; namely, he records [122] (as early as in 1933)

$$M(x^{10} + x^9 - x^7 - x^6 - x^5 - x^4 - x^3 + x + 1) = 1.17628081\ldots . \qquad (1.2)$$

This still stands as the smallest value of $M(P) > 1$, in spite of extensive computation done since 1933 by many mathematicians. At the same time, the problem itself remains open.

Here we give a simple argument on the bound of $M(P)$ from below in the case that the degree of $P(x)$ is bounded.

Lemma 1.3 *For every $d \in \mathbb{Z}_{>0}$ there exists $\mu_d > 1$ such that if $P(x) \in \mathbb{Z}[x]$ is a monic, irreducible and non-cyclotomic polynomial of degree d, then $M(P) \geq \mu_d$.*

Proof Since $M(x^d - 2) = 2$, it is sufficient to show that there are *finitely* many monic polynomials $P(x) \in \mathbb{Z}[x]$ of degree d for which $M(P) < 2$. Then μ_d is the minimum of $M(P)$ over the finite set.

For a polynomial $P(x) = \prod_{j=1}^{d}(x - \alpha_j)$, the bound $M(P) = \prod_{1 \leq j \leq d} \max\{1, |\alpha_j|\} < 2$ implies $|\alpha_j| < 2$ for $j = 1, \ldots, d$. As the *integral* coefficients of $P(x)$ are elementary symmetric polynomials of $\alpha_1, \ldots, \alpha_d$, the estimates impose certain bounds on the absolute value of the coefficients of $P(x)$. This implies that the set of such polynomials is indeed finite. $\square$

We conclude this section with a problem related to another classical sequence of Fermat primes — primes of the form $2^{2^n} + 1$, $n = 0, 1, 2, \ldots$.

Exercise 1.8 ([155, Division 8, Chapter 2, Problem 94]) Show that all terms of the sequence

$$2^1 + 1, \ 2^2 + 1, \ 2^4 + 1, \ 2^8 + 1, \ \ldots, \ 2^{2^n} + 1, \ \ldots$$

are pairwise coprime.

In fact, this exercise shows that the number of primes that do not exceed a given $x \geq 3$, is at least $c \log \log x$ for an absolute constant $c > 0$.

1.3 Jensen's formula

Proposition 1.4 (Jensen's formula [107]) *The following evaluation takes place:*

$$\int_0^1 \log|e^{2\pi it} - \alpha|\, dt = \log^+ |\alpha|,$$

where $\log^+ |\alpha| = \max\{0, \log|\alpha|\} = \log\max\{1, |\alpha|\}$.

The result is trivially true for $\alpha = 0$ (when the logarithm is supposed to be $-\infty$), so that we assume $|\alpha| > 0$ below.

Complex-analytic proof   Changing to the complex variable $x = e^{2\pi it}$,

$$\int_0^1 \log|e^{2\pi it} - \alpha|\, dt = \frac{1}{2\pi i} \oint_{|x|=1} \log|x - \alpha| \frac{dx}{x} = \mathrm{Re}\left(\frac{1}{2\pi i} \oint_{|x|=1} \log(x - \alpha) \frac{dx}{x}\right).$$

If $|\alpha| > 1$, then $f(x) = \log(x - \alpha)$ is analytic inside the unit disc $|x| < 1$ and on its boundary, hence Cauchy's theorem implies that the latter integral evaluates to $f(0) = \mathrm{Re}\log\alpha = \log|\alpha|$.

If $|\alpha| = 1$, so that the integral is improper, we replace the integration path around $x = \alpha$ with an arc C of radius ε, where $0 < \varepsilon < \frac{1}{2}$, centred at this point lying entirely inside the disc $|x| \le 1$, and use Cauchy's theorem for the newer contour as well as the estimate

$$\left|\frac{1}{2\pi i} \int_C \log(x - \alpha) \frac{dx}{x}\right| \le \varepsilon \max_{x:|x-\alpha|=\varepsilon} \frac{|\log(x-\alpha)|}{|x|} \le 2\varepsilon|\log\varepsilon| \to 0 \quad \text{as } \varepsilon \to 0.$$

Finally, if $|\alpha| < 1$, then $\log|x-\alpha| = \log|1-\alpha/x|$ on the contour of integration $|x| = 1$, so that

$$\int_0^1 \log|e^{2\pi it} - \alpha|\, dt = \mathrm{Re}\left(\frac{1}{2\pi i} \oint_{|x|=1} \log\left(1 - \frac{\alpha}{x}\right) \frac{dx}{x}\right)$$

$$= \mathrm{Re}\left(\frac{1}{2\pi i} \oint_{|y|=1} \log(1 - \alpha y) \frac{dy}{y}\right).$$

It remains to observe that the integrand $y^{-1}\log(1 - \alpha y)$ has a removable singularity at $y = 0$ and no other within the disc $|y| \le 1$, implying that the integral indeed evaluates to 0.   □

Real-analytic proof   Write $re^{2\pi is}$ for α and shift t by s to reduce the integral under consideration to

$$\int_0^1 \log|e^{2\pi it} - \alpha|\, dt = \int_0^1 \log|e^{2\pi it} - r|\, dt,$$

where $r = |\alpha| > 0$. The resulting integral is seen to be the limit of the integral sums,

$$\int_0^1 \log |e^{2\pi i t} - r| \, dt = \lim_{n \to \infty} \frac{1}{n} \sum_{k=1}^{n} \log |e^{2\pi i k/n} - r|$$

$$= \lim_{n \to \infty} \frac{1}{n} \log \left| \prod_{k=1}^{n} (e^{2\pi i k/n} - r) \right| = \lim_{n \to \infty} \frac{\log |1 - r^n|}{n},$$

and the desired formula follows. □

Exercise 1.9 Show the 'real-looking' version of Jensen's formula:

$$\frac{1}{2} \int_0^1 \log(1 - 2r \cos 2\pi t + r^2) \, dt = \log^+ r.$$

Hint Notice that $1 - 2r \cos 2\pi t + r^2 = |e^{2\pi i t} - r|^2$. □

Let us follow Mahler's steps [129] and replace the linear polynomial $x - \alpha$ with a general polynomial $P(x) = a_d \prod_{j=1}^{d} (x - \alpha_j)$:

$$\int_0^1 \log |P(e^{2\pi i t})| \, dt = \frac{1}{2\pi i} \oint_{|x|=1} \log |a_d| \frac{dx}{x} + \sum_{j=1}^{d} \frac{1}{2\pi i} \oint_{|x|=1} \log |x - \alpha_j| \frac{dx}{x}$$

$$= \log |a_d| + \sum_{j=1}^{d} \log \max\{1, |\alpha_j|\}.$$

Proposition 1.5 ([129]) *For a polynomial* $P(x) = a_d \prod_{j=1}^{d} (x - \alpha_j)$,

$$\int_0^1 \log |P(e^{2\pi i t})| \, dt = \frac{1}{2\pi i} \oint_{|x|=1} \log |P(x)| \frac{dx}{x} = \log |a_d| + \sum_{j=1}^{d} \log^+ |\alpha_j|.$$

Comparing this result to the characteristic witnessed in Theorem 1.2, we define the Mahler measure of a polynomial $P(x) = a_d \prod_{j=1}^{d} (x - \alpha_j) \in \mathbb{C}[x]$ as

$$M(P) = |a_d| \prod_{j=1}^{d} \max\{1, |\alpha_j|\}. \tag{1.3}$$

Then

$$M(P) = \exp\left(\int_0^1 \log |P(e^{2\pi i t})| \, dt \right) = \exp\left(\frac{1}{2\pi i} \oint_{|x|=1} \log |P(x)| \frac{dx}{x} \right) \tag{1.4}$$

is the Mahler measure of the polynomial $P(x)$, while the integrals in Proposition 1.5 represent the logarithmic Mahler measure

$$m(P) = \log M(P) = \int_0^1 \log |P(e^{2\pi i t})| \, dt = \frac{1}{2\pi i} \oint_{|x|=1} \log |P(x)| \frac{dx}{x}.$$

The adjective 'logarithmic' is often dropped when use of m(·) rather than of M(·) is clear from the context.

Note that the definition implies $M(PQ) = M(P) \cdot M(Q)$ for polynomials $P(x)$ and $Q(x)$ — in other words, the Mahler measure is multiplicative. We can even extend it unambiguously from polynomials to rational functions by setting $M(P/Q) = M(P)/M(Q)$.

For polynomials $P(x)$ with *integer* coefficients, clearly $M(P) \geq 1$ with $M(P) = 1$ only if P is monic ($a_d = 1$) and has all its zeros inside the unit circle (hence is a product of a monomial x^n and a cyclotomic polynomial, by Kronecker's theorem).

Exercise 1.10 Show that if $P(x) \in \mathbb{Z}[x]$ then its Mahler measure $M(P)$ is an algebraic *integer*.

Lemma 1.6 *For a polynomial* $P(x) = a_d x^d + \cdots + a_1 x + a_0 \in \mathbb{C}[x]$,

$$|a_n| \leq \binom{d}{n} M(P), \quad \textit{where } n = 0, 1, \ldots, d.$$

Proof According to Viète's theorem, the coefficient a_{d-n} of $P(x)$ is, up to sign, the sum of terms of the form $a_d \alpha_{j_1} \cdots \alpha_{j_n}$, where $\alpha_1, \ldots, \alpha_d$ are the zeros (counted with multiplicity) of $P(x)$. It readily follows from (1.3) that each such term is bounded from above by $M(P)$, and the result then follows from the fact that the number of terms is exactly $\binom{d}{d-n} = \binom{d}{n}$.   □

Lemma 1.7 *For a polynomial* $P(x) = a_d x^d + \cdots + a_1 x + a_0 \in \mathbb{C}[x]$,

$$M(P) \leq L(P) = |a_0| + |a_1| + \cdots + |a_d|, \quad \textit{the length of } P(x).$$

Proof The estimate follows from Proposition 1.5 and the trivial estimate $|P(x)| \leq L(P)$ for x on the unit circle, $|x| = 1$.   □

Exercise 1.11 (Mignotte [139]) Define the norm of polynomial $P(x) = a_d x^d + \cdots + a_1 x + a_0 \in \mathbb{C}[x]$ by

$$\|P(x)\| = (|a_d|^2 + \cdots + |a_1|^2 + |a_0|^2)^{1/2} = \left(\int_0^1 |P(e^{2\pi i t})|^2 dt \right)^{1/2}.$$

(a) Show that $M(P) \leq \|P(x)\|$.
(b) If $P(x) = a_d x^d + \cdots + a_1 x + a_0 \in \mathbb{Z}[x]$ divides a polynomial $Q(x)$ with *integer* coefficients then

$$|a_n| \leq \binom{d}{n} \|Q(x)\| \quad \text{for } n = 0, 1, \ldots, d.$$

The latter bounds, known as the Landau–Mignotte bounds, allow one to search for potential factors of a given polynomial $Q(x) \in \mathbb{Z}[x]$.

Proposition 1.8 (Gelfond's inequality) *For a product* $P(x) = P_1(x) \cdots P_m(x)$ *of polynomials, we have*

$$L(P) \leq L(P_1) \cdots L(P_m) \leq 2^{\deg P} L(P).$$

Proof The lower bound for $L(P_1) \cdots L(P_m)$ is nearly trivial: it does not involve the Mahler measure at all.

In the other direction, it follows from Lemma 1.6 that, for a polynomial $P(x) = a_d x^d + \cdots + a_1 x + a_0$,

$$L(P) = \sum_{n=0}^{d} |a_n| \leq \sum_{n=0}^{d} \binom{d}{n} M(P) = 2^d M(P).$$

Writing this inequality for each factor $P_j(x)$ and using $\sum_{j=1}^{m} \deg P_j = \deg P$, $M(P_1 \cdots P_m) = M(P)$ and the bound from Lemma 1.7, we arrive at the required claim. $\qquad\square$

Finally, we define the Mahler measure and logarithmic Mahler measure of an algebraic number α as the corresponding quantities for the minimal polynomial $P(x) \in \mathbb{Z}[x]$ of α. In particular, $M(\alpha) \geq 1$ for an algebraic number α with the equality happening only when $\alpha = 0$ or $\alpha^n = 1$ for some $n \in \mathbb{Z}_{>0}$.

Exercise 1.12 The house $\lceil \alpha \rceil$ of an algebraic integer α of degree d is defined as the maximum modulus of its conjugates (including α itself).

(a) Show that $M(\alpha)^{1/d} \leq \lceil \alpha \rceil \leq M(\alpha)$.
(b) If $M(\alpha) < 2$, prove a refined estimate $M(\alpha) \leq (\max\{\lceil \alpha \rceil, \lceil 1/\alpha \rceil\})^{d/2}$.

Hint (b) Observe that in this case $M(\alpha) = M(\alpha^{-1})$. $\qquad\square$

1.4 Families of Mahler measures

Before going on with partial resolutions of Lehmer's problem, we treat the 'baby' families

$$\lambda^-(k) = m\left(x - \frac{1}{x} + k\right) = m(x^2 + kx - 1) = \log \frac{k + \sqrt{k^2 + 4}}{2}$$

and

$$\lambda^+(k) = m\left(x + \frac{1}{x} + k\right) = m(x^2 + kx + 1) = \log \frac{k + \sqrt{k^2 - 4}}{2},$$

where k is a positive integer (and $k \geq 3$ in the latter case), and their classical links to the arithmetic of real quadratic fields. Observe that the numbers

$$\varepsilon_k^- = \frac{k + \sqrt{k^2 + 4}}{2} \quad \text{and} \quad \varepsilon_k^+ = \frac{k + \sqrt{k^2 - 4}}{2}$$

are units in the corresponding real quadratic fields $\mathbb{Q}(\sqrt{k^2 + 4})$ and $\mathbb{Q}(\sqrt{k^2 - 4})$, because their norms are equal to ± 1 (a consequence of the underlying quadratic polynomials). In fact, the purely periodic continued fraction

$$\varepsilon_k^- = k + \cfrac{1}{k + \cfrac{1}{k + \cfrac{1}{\ddots}}}$$

for the first unit implies that $\varepsilon_k^- > 1$ is the fundamental unit of $\mathbb{Q}(\sqrt{k^2 + 4})$ for odd $k > 0$ (see, for example, [33, Chapter 4]).

The structure of a real quadratic field $K = \mathbb{Q}(\sqrt{D})$ is particularly simple: the regulator is given by the logarithm of its fundamental unit $\varepsilon_D > 1$, and the latter generates (up to sign) the multiplicative group of units of K. Applied to our situations, this leads to the statements

$$\lambda^-(k) = r_k^- \cdot \operatorname{Reg} \mathbb{Q}(\sqrt{k^2 + 4}) \quad \text{and} \quad \lambda^+(k) = r_k^+ \cdot \operatorname{Reg} \mathbb{Q}(\sqrt{k^2 - 4})$$

for some positive integers r_k^- and r_k^+; more specifically, $r_k^- = 1$ for k odd.

Now let D be the fundamental discriminant of the field $K = \mathbb{Q}(\sqrt{k^2 + 4})$ or $\mathbb{Q}(\sqrt{k^2 - 4})$, so that $D = k^2 + 4$ or $k^2 - 4$, except for the situation when $k \equiv 2 \bmod 4$ in the second case and D is $(k^2 - 4)/4^\ell$ for ℓ such that $(k^2 - 4)/4^{\ell+1}$ is not divisible by 4. Denote by $\chi(n) = \chi_D(n) = \left(\frac{D}{n}\right)$ the corresponding Dirichlet character, where $\left(\frac{D}{\cdot}\right)$ is the Kronecker symbol. Then the Dirichlet class number formula for the field K asserts that

$$\sqrt{D} \, L(\chi, 1) = h(D) \log \varepsilon_D,$$

where $h(D)$ is the number of equivalence classes of quadratic forms with discriminant D and

$$L(\chi, s) = \sum_{n=1}^{\infty} \frac{\chi(n)}{n^s}$$

stands for the Dirichlet L-function. Restricted to our situations, this gives

$$\lambda^-(k) = \tilde{r}_k^- \cdot \sqrt{k^2 + 4} \, L(\chi, 1) \quad \text{and} \quad \lambda^+(k) = \tilde{r}_k^+ \cdot \sqrt{k^2 - 4} \, L(\chi, 1)$$

for some positive rationals $\tilde{r}_k^-$ and $\tilde{r}_k^+$.

The fact that the quantities $\sqrt{D} \, L(\chi_D, 1)$ and $\operatorname{Reg} K$, for a real quadratic field $K = \mathbb{Q}(\sqrt{D})$, are rationally proportional to each other is classical and follows from the Dirichlet class number formula. Perhaps it is more surprising that the two characteristics are also proportional to the Mahler measure of the corresponding quadratic polynomial.

Exercise 1.13 Show that any real quadratic field K can be represented as $\mathbb{Q}(\sqrt{k^2 - 4})$ for some integer $k \geq 3$.

Hint Let $D > 0$ be the fundamental discriminant of K. Then Pell's equation $X^2 - DY^2 = 4$ always has a solution in positive X, Y (in fact, its minimal solution $X_0 \geq Y_0 > 0$ corresponds to the unit $\frac{1}{2}(X_0 + Y_0 \sqrt{D})$ which is either fundamental or the square of the fundamental unit). Take $k = X$. $\qquad\square$

Chapter notes

Jensen's motivation for writing the manuscript [107] was the hope of proving the Riemann hypothesis (more details of his approach can be found in [95]). Though the latter did not materialise, the discovered formula (Proposition 1.4) continues to find applications throughout mathematics.

One thing we can learn from the real-analytic proof of Jensen's formula in Section 1.3 is the form

$$\mathrm{m}(P) = \lim_{n \to \infty} \frac{1}{n} \sum_{k \bmod n} \log |P(\zeta_n^k)| \tag{1.5}$$

for computing the logarithmic Mahler measure of a polynomial $P(x)$. This is different from the analytic expression in the section but somewhat original in the works of Pierce [152] and Lehmer [122]; the form nicely serves as a source for a discrete version of the Mahler measure and analogues of Lehmer's problem for the latter [123, 153]. In fact, the existence of the integral in Proposition 1.5 implies that its value can be realised as a limit of regular (that is, equally sized) Riemann sums, which (in view of the 1-periodicity of $P(e^{2\pi i t})$) indeed leads to the form (1.5). Later, in Section 3.4, we will witness the usefulness of the expression.

In general, the regulator of a number field K is an $r \times r$ determinant whose entries are the logarithms of units in K, where r is the rank of the group of units. This means that the regulator can be *potentially* compared to the logarithmic Mahler measure of a polynomial $P(x) \in \mathbb{Z}[x]$ whose Galois extension is K only if $r = 1$. Apart from our examples in Section 1.4 this also appears for certain cubic polynomials $x^3 + ax^2 + bx - 1$, namely, for those having two complex conjugate zeros inside the unit disc and a single real zero greater than 1. For example, the real zero $\alpha = \alpha(k) > 1$ of $x^3 - kx^2 - 1$, where $k \in \mathbb{Z}_{>0}$, is the fundamental unit of $K = \mathbb{Q}(\alpha)$ [186, Theorem 1.1]; thus, the regulator of K is rationally proportional to $\log \alpha = m(x^3 - kx^2 - 1)$. Then the class number formula relates the former to the residue of the Dedekind zeta function $\zeta_K(s)$ at $s = 1$.

Additional exercises

Exercise 1.14 ([157, Section 13]) (a) For p prime, show that

$$\Phi_{np}(x) = \begin{cases} \Phi_n(x^p) & \text{if } (n, p) = p, \\ \Phi_n(x^p)/\Phi_n(x) & \text{if } (n, p) = 1. \end{cases}$$

(b) Using part (a) or otherwise show that, for $n > 1$, one has

$$\Phi_n(1) = \begin{cases} p & \text{if } n = p^m \text{ for some } m \geq 1, \\ 1 & \text{otherwise.} \end{cases}$$

Give a similar formula for $\Phi_n(-1)$.

Exercise 1.15 (Suzuki [205]; see also [157, Section 13], [108]) (a) What   is the smallest index n for which the maximum of the absolute values of the coefficients of $\Phi_n(x)$ is greater than 1? Greater than 2?

(b) Prove that any integer m can occur as a coefficient of some cyclotomic polynomial $\Phi_n(x)$.

Exercise 1.16 (Gauss's cyclotomic formula [90, 113]) Show that, for odd square-free $n > 3$, the cyclotomic polynomial $\Phi_n(x)$ can be represented in the form

$$\Phi_n(x) = \frac{1}{4}(A_n^2(x) - (-1)^{(n-1)/2} n x^2 B_n^2(x)),$$

where the polynomials $A_n(x)$ and $B_n(x)$ (of degree $\varphi(n)/2$ and $\varphi(n)/2 - 1$, respectively) have integral coefficients and satisfy $A_n^*(x) = (-1)^{(n-1)/2} A_n(x)$ and $B_n^*(x) = B_n(x)$ (in other words, $B_n(x)$ is reciprocal — check the notation in Section 1.1).

Hint   The wanted representation is equivalent to

$$4\Phi_n(x) = (A_n(x) - (-1)^{(n-1)/4} \sqrt{n} x B_n(x))(A_n(x) + (-1)^{(n-1)/4} \sqrt{n} x B_n(x)).$$

Design a splitting of the $\varphi(n)$ zeros $\zeta_n^k = e^{2\pi i k/n}$ of the polynomial on the left-hand side into two disjoint sets Z and $\overline{Z}$ in such a way that

$$2 \prod_{\zeta \in Z} (x - \zeta) \in \mathbb{Z}[x] + (-1)^{(n-1)/4} \sqrt{n} \mathbb{Z}[x]. \qquad \square$$

Exercise 1.17 (Lucas' cyclotomic formula [113]) Show that, for a square-free integer $n \geq 3$, the cyclotomic polynomial $\Phi_n(x)$ can be represented in the form

$$\Phi_n(x) = C_n^2(x) - (-1)^{(n-1)/2} nx D_n^2(x),$$

where the polynomials $C_n(x)$ and $D_n(x)$ (of degree $\varphi(n)/2$ and $\varphi(n)/2 - 1$, respectively) have integral coefficients and satisfy $C_n^*(x) = \pm C_n(x)$ and $D_n^*(x) = \pm D_n(x)$.

Hint Substitute x^2 for x in the wanted formula and write it as

$$\Phi_n(x^2) = (C_n(x^2) - (-1)^{(n-1)/4} \sqrt{n} x D_n(x^2))(C_n(x^2) + (-1)^{(n-1)/4} \sqrt{n} x D_n(x^2)).$$

Then split the $2\varphi(n)$ zeros $\{e^{\pi ik/n} : (k, n) = 1, \ k = 1, 2, \dots, 2n\}$ into two disjoint sets in a way similar to the previous exercise. □

Numerical examples illustrating Exercises 1.16 and 1.17 can be found in [163, Tables 23, 24].

Exercise 1.18 (Pomerance, Rubinstein-Salzedo [156]) Show that if $\Phi_n(q) = \Phi_m(q)$ then either $n = m$, or $1/2 < q < 2$, or $\Phi_2(2) = \Phi_6(2)$.

In particular, this implies the cyclotomic ordering conjecture [94]: for any pair m, n of positive integers, either $\Phi_m(q) \leq \Phi_n(q)$ holds for all integers $q \geq 2$ or the reverse inequality holds for all such q.

2

Lehmer's problem

In this chapter we discuss resolved cases of and results related to Lehmer's problem — finding the minimal size of the Mahler measure of a non-cyclotomic polynomial $P(x) \in \mathbb{Z}[x]$, equivalently, of an algebraic number α which is not a root of unity.

2.1 Smyth's theorem

Recall that a polynomial $P(x)$ of degree d is called reciprocal if $P(x) = P^*(x)$, where $P^*(x) = x^d P(1/x)$. The following characterisation of Mahler measures of non-reciprocal polynomials was given by Smyth [195, 196] in 1971.

Theorem 2.1 (Smyth [195]) *If the Mahler measure of an irreducible polynomial $P(x) \in \mathbb{Z}[x]$ satisfies $\mathrm{M}(P) < x_0 = 1.32471795\ldots$, the real zero of the polynomial $x^3 - x - 1$, then $P(x)$ is reciprocal.*

Furthermore, if $\mathrm{M}(P) \leq c = x_0 + 10^{-4}$, then either $P(x)$ has a zero $x_0^{1/m}$ for some $m \in \mathbb{Z}_{>0}$ or $P(x)$ is reciprocal.

In fact, the constant $c = x_0 + 10^{-4}$ in the statement was later improved to $c = 1.32497826\ldots$, the largest real zero of $4x^8 - 5x^6 - 2x^4 - 5x^2 + 4$, in [74, Theorem 15] using related estimates from [196].

In this section we present a slightly weaker version of the theorem.

Theorem 2.2 *If the Mahler measure of an irreducible polynomial $P(x) \in \mathbb{Z}[x]$ satisfies $\mathrm{M}(P) \leq c = (1 + \sqrt{17})/4 = 1.28077640\ldots$, then $P(x)$ is reciprocal.*

In 1951, Breusch [46] gave a version of the result with constant $c = x_0 = 1.17965204\ldots$, the zero of $x^3 - x^2 - \frac{1}{4}$.

The proof of Theorem 2.2 below closely follows the original publication [195] and its exposition in [176, Chapter 6]; the method of proof goes back

14

to Salem's 1944 work [171] (see also Siegel's follow-up in [189]). Note that Smyth's theorem (also in its simplified form of Theorem 2.2) gives another solution to Exercise 1.4.

Lemma 2.3 *Assume that $f(z) = \sum_{n=0}^{\infty} a_n z^n$ is a series expansion of a non-constant function that is analytic inside the unit disc, continuous on its closure and satisfies $|f(z)| \leq 1$ for $|z| \leq 1$. Then*

$$|a_0| < 1, \quad |a_1| \leq 1 - |a_0|^2.$$

Proof The fact that $|a_0| = |f(0)| \leq 1$ immediately follows from the maximum principle. For an analytic function $f(z)$ on an open domain $\mathcal{D}$ (which we assume here to be the unit disc $\{z : |z| < 1\}$), continuous on its closure $\overline{\mathcal{D}}$ and bounded by a positive constant M for $z \in \overline{\mathcal{D}}$, one has the strict inequality $|f(z)| < M$ for $z \in \mathcal{D}$ unless $f(z)$ is a constant. Under the hypothesis of the lemma, we get $|f(z)| < 1$ for $|z| < 1$; in particular, $|a_0| = |f(0)| < 1$.

When $f(0) = 0$, the same principle applied to $f(z)/z$ shows that if $|f(z)| \leq 1$ for $|z| \leq 1$, then $|f(z)| \leq |z|$ for $|z| \leq 1$ (and even a strict inequality for $|z| < 1$ provided $f(z)$ is not linear); this is known as Schwarz's lemma.

For any a, $|a| < 1$, the mapping $w \mapsto (w - a)/(1 - \bar{a}w)$ bijectively maps the unit disc $\{w : |w| < 1\}$ into itself; this means that the function

$$g(z) = \frac{f(z) - a_0}{1 - \bar{a}_0 f(z)}$$

satisfies $|g(z)| \leq 1$ for $|z| \leq 1$. By Schwarz's lemma, $|g(z)| \leq |z|$, so that

$$\left| \frac{f(z) - a_0}{z} \right| \leq |1 - \bar{a}_0 f(z)| \quad \text{for } 0 < |z| \leq 1,$$

and the limiting case, as $z \to 0$, reads $|a_1| \leq |1 - \bar{a}_0 a_0| = |1 - |a_0|^2| = 1 - |a_0|^2$. $\square$

Exercise 2.1 (Carathéodory [57], [154, Division 3, Chapter 6, Problem 282]) Under the hypothesis of the lemma show that

$$|f(z) - f(0)| \leq |z| \frac{1 - |f(0)|^2}{1 - |f(0)| \, |z|} \quad \text{for } |z| \leq 1.$$

This inequality is strict for $0 < |z| < 1$ unless

$$f(z) = \frac{e^{i\theta} z - z_0}{1 - \bar{z}_0 e^{i\theta} z}$$

for some real θ.

Lemma 2.4 *Assume that $f(z) = \sum_{n=0}^{\infty} a_n z^n$ is a series expansion of a function that is analytic inside the unit disc, continuous on its closure and satisfies*

$|f(z)| \le 1$ *for* $|z| \le 1$. *Then, for each* $k = 1, 2, \ldots,$ *the 'rarefied' series expansion* $f_k(z) = \sum_{n=0}^{\infty} a_{nk} z^n$ *features the same properties.*

In particular, the estimate of Lemma 2.3 extends to $|a_k| \le 1 - |a_0|^2$ *for any* $k \in \mathbb{Z}_{>0}$.

Proof As

$$f_k(z^k) = \frac{1}{k} \sum_{j=1}^{k} f(e^{2\pi i j/k} z),$$

the statement follows immediately from

$$|f_k(z^k)| \le \frac{1}{k} \sum_{j=1}^{k} |f(e^{2\pi i j/k} z)| \le 1 \quad \text{for } |z| < 1. \qquad \square$$

Proof of Theorem 2.2   Suppose, on the contrary, that there exists a non-reciprocal irreducible polynomial $P(x) = a \prod_{j=1}^{d} (x - \alpha_j) \in \mathbb{Z}[x]$ whose Mahler measure

$$\mu = \mathrm{M}(P) = |a| \prod_{j=1}^{d} \max\{1, |\alpha_j|\} = |a| \prod_{j:|\alpha_j|>1} |\alpha_j| \le c = \frac{1 + \sqrt{17}}{4}.$$

This clearly implies that $|a| \le c < 2$, so that we assume $a = 1$. This also means that $|P(0)| = \prod_{j=1}^{d} |\alpha_j| \le \mu < 2$, hence $|P(0)| = 1$, so that the polynomial $P^*(x) = x^d P(1/x)$ is monic as well. Because $P^*(x)$ differs from $P(x)$, the rational function $P(x)/P^*(x)$ is not constant: we cannot have $P(x) = -P^*(x)$ either, as then $P(1) = -P^*(1) = -P(1)$ implying $P(1) = 0$, which is impossible.

As $P(0)P(z) = 1 + p_1 z + \cdots$ and $P^*(z) = 1 + p_1^* z + \cdots$ are power series with integral rational coefficients, so is the rational function

$$\frac{P(0)P(z)}{P^*(z)} = 1 + c_1 z + c_2 z^2 + \cdots \in \mathbb{Z}[[z]].$$

We pick up the first index $k \ge 1$ for which the coefficient c_k is non-zero:

$$\frac{P(0)P(z)}{P^*(z)} = 1 + c_k z^k + \cdots \in \mathbb{Z}[[z]]. \tag{2.1}$$

Furthermore, we have

$$P(z) = \prod_{j=1}^{d} (z - \alpha_j) = \prod_{j=1}^{d} (z - \bar{\alpha}_j) \quad \text{and} \quad P^*(z) = \prod_{j=1}^{d} (1 - \alpha_j z) = \prod_{j=1}^{d} (1 - \bar{\alpha}_j z).$$

Splitting the zeros $\{\alpha_j : j = 1, \ldots, d\} = \{\bar{\alpha}_j : j = 1, \ldots, d\}$ of $P(z)$ into three disjoint sets (each stable under complex conjugation!), according to whether $|\alpha_j| < 1$, $|\alpha_j| = 1$ or $|\alpha_j| > 1$, write

$$\frac{P(0)P(z)}{P^*(z)} = \frac{f(z)}{g(z)},$$

where

$$f(z) = \pm \prod_{|\alpha_j|<1} \frac{z - \alpha_j}{1 - \bar{\alpha}_j z} \quad \text{and} \quad g(z) = \pm \prod_{|\alpha_j|>1} \frac{1 - \bar{\alpha}_j z}{z - \alpha_j};$$

the signs are chosen to achieve $f(0) > 0$ and $g(0) > 0$. As already seen in the proof of Lemma 2.3, the absolute values of each factor in the products for $f(z)$ and $g(z)$ are bounded from above by 1 for $|z| \leq 1$. This implies that the series expansions of the rational functions

$$f(z) = a_0 + a_1 z + a_2 z^2 + \cdots \in \mathbb{R}[[z]] \quad \text{and} \quad g(z) = b_0 + b_1 z + b_2 z^2 + \cdots \in \mathbb{R}[[z]]$$

satisfy the conditions of Lemmas 2.3 and 2.4. It follows from their definitions that $b_0 = g(0) = \mu^{-1}$. On comparing the series expansions with (2.1) we deduce that

$$a_0 = b_0 = \mu^{-1}, \quad a_1 = b_1, \quad \ldots, \quad a_{k-1} = b_{k-1} \text{ and } a_k = b_k + c_k b_0.$$

Recall that c_k in (2.1) is a non-zero integer, $|c_k| \geq 1$; it follows from the last relation that $\max\{|a_k|, |b_k|\} \geq |c_k|b_0/2 \geq b_0/2 = \frac{1}{2}\mu^{-1}$. On the other hand, Lemma 2.4 implies $\max\{|a_k|, |b_k|\} \leq 1 - \mu^{-2}$. The inequality $\frac{1}{2}\mu^{-1} \leq \max\{|a_k|, |b_k|\} \leq 1 - \mu^{-2}$ implies $\mu \geq (1 + \sqrt{17})/4$. Furthermore, this last inequality must be strict, because μ happens to be an algebraic *integer*. The contradiction thus shows that the polynomial $P(x)$ must be reciprocal. □

In the remainder of this section we give two more special results on estimates of the Mahler measure from below: for the case of totally real algebraic integers (Schinzel's theorem [175, 101]) — Exercise 2.3, and for $M(\alpha) M(1-\alpha)$ (a theorem by Zhang [220] and its refinement by Zagier [218]) — Exercise 2.4.

Exercise 2.2 (a) For all $z \in \mathbb{R} \setminus \{0\}$, show that

$$\max\{0, \log|z|\} \geq \frac{\sqrt{5} - 1}{2\sqrt{5}} \log|z| + \frac{1}{2\sqrt{5}} \log|z^2 - 1| + \frac{1}{2} \log \frac{\sqrt{5} + 1}{2},$$

with equality if and only if z equals $\pm(\sqrt{5} \pm 1)/2$.
(b) For all $z \in \mathbb{C} \setminus \{0, 1, (1 \pm i\sqrt{3})/2\}$, show that

$$\max\{0, \log|z|\} + \max\{0, \log|1 - z|\}$$

$$\geq \frac{\sqrt{5} - 1}{2\sqrt{5}} \log|z^2 - z| + \frac{1}{2\sqrt{5}} \log|z^2 - z + 1| + \frac{1}{2} \log \frac{\sqrt{5} + 1}{2},$$

with equality if and only if z or $1 - z$ equals $e^{\pm\pi i/5}$ or $e^{\pm 3\pi i/5}$.

Solution   (a) We follow the lemma in [100] and, for real a in the interval $0 < a < \frac{1}{2}$, consider the real-valued (non-negative) function

$$g_a(z) = \frac{|z|^{1/2}|z - 1/z|^a}{\max\{1, |z|\}}, \quad \text{where } z \in \mathbb{C} \setminus \{0\}.$$

Notice that $g_a(1/z) = g_a(z) = g_a(-z)$. Therefore,

$$\max_{z \in \mathbb{R} \setminus \{0\}} g_a(z) = \max_{0 < z < 1} g_a(z) = g_a\left(\sqrt{\frac{1 - 2a}{1 + 2a}}\right) = G(a)$$

with $G(a) = (4a)^a (1 - 2a)^{(1-2a)/4}(1 + 2a)^{-(1+2a)/4}$. (One also has

$$\max_{z \in \mathbb{C} \setminus \{0\}} g_a(z) = \max_{|z|=1} g_a(z) = g_a(i) = 2^a,$$

so that $g_a(z) \le 2^a$ for all complex z.) The required estimate now follows from taking $a = 1/(2\sqrt{5})$ — the value that minimises $G(a)$ on the interval $0 < a < \frac{1}{2}$.

(b) This is quite similar to part (a); the details can be found in [218].   □

Exercise 2.3 (Schinzel [175]) For a totally real algebraic integer α of degree $d \ge 2$, show that

$$M(\alpha) \ge \left(\frac{1 + \sqrt{5}}{2}\right)^{d/2}.$$

Solution [101]   For the minimal polynomial $P(x) = \prod_{j=1}^{d}(x - \alpha_j)$ of α, we have $|P(0)| \ge 1$, $|P(1)| \ge 1$ and $|P(-1)| \ge 1$. Therefore, introducing $f(x) = |x|^{1/2}|x - 1/x|^{1/(2\sqrt{5})}$, we obtain

$$\prod_{j=1}^{d} f(\alpha_j) = |P(0)|^{1/2 - 1/(2\sqrt{5})}|P(1)P(-1)|^{1/(2\sqrt{5})} \ge 1.$$

On the other hand, Exercise 2.2(a) implies that

$$f(\beta) \le \left(\frac{1 + \sqrt{5}}{2}\right)^{-1/2} \max\{1, |\beta|\}$$

for real $\beta \ne 0$, and the desired result follows on combining the two estimates.

□

Exercise 2.4 (Zhang [220], Zagier [218]) For any algebraic number α different from $0, 1, (1 \pm i\sqrt{3})/2$, show that

$$M(\alpha)\,M(1 - \alpha) \ge \left(\frac{1 + \sqrt{5}}{2}\right)^{1/2},$$

with equality if and only if α or $1 - \alpha$ is a primitive 10th root of unity.

Hint   Use Exercise 2.2(b).   □

2.2 Dobrowolski's bound

In this section we discuss a remarkable achievement in the resolution of Lehmer's problem for *any* algebraic α different from a root of unity (equivalently, for any irreducible non-cyclotomic polynomial $P(x) \in \mathbb{Z}[x]$).

Theorem 2.5 (Dobrowolski [75]) *For any algebraic number α of degree d, which is not a root of unity,*

$$\mathrm{m}(\alpha) \geq c_0 \left(\frac{\log \log d}{\log d} \right)^3 \tag{2.2}$$

with some absolute constant $c_0 > 0$.

The constant was $1 - \varepsilon$ in Dobrowolski's original proof [75], where $\varepsilon > 0$ is arbitrary and $d > d_0(\varepsilon)$ is sufficiently large; he also offered the choice $\frac{1}{1200}$ valid for any d. The later proof given by Cantor and Straus [56] has the better constant $2 - \varepsilon$ for c_0, which was further improved to $\frac{9}{4} - \varepsilon$ by Louboutin [127] (see also [176, Section 4.2]).

In our proof of Theorem 2.5 we follow closely the exposition in [56], which makes use of the following beautiful determinant.

Lemma 2.6 (Confluent Vandermonde determinant [114, Theorem 20]) *For $x \in \mathbb{C}$, define the $n \times m$ matrix $A_{n,m}(x)$ as*

$$\begin{pmatrix}
1 & 0 & 0 & 0 & \cdots & 0 \\
x & 1 & 0 & 0 & \cdots & 0 \\
x^2 & 2x & 1 & 0 & \cdots & 0 \\
x^3 & 3x^2 & 3x & 1 & \cdots & 0 \\
\cdots & \cdots & \cdots & \cdots & \cdots & \cdots \\
x^{n-1} & (n-1)x^{n-2} & \binom{n-1}{2}x^{n-3} & \cdots & \binom{n-1}{m-2}x^{n-m+1} & \binom{n-1}{m-1}x^{n-m}
\end{pmatrix}$$

so that the kth column, $k = 2, \ldots, m$ is the result of application of the differential operator $\frac{1}{(k-1)!} \frac{d^k}{dx^k}$ to the first one. Given a composition $n = m_1 + m_2 + \cdots + m_r$, we have

$$\det(A_{n,m_1}(x_1) \; A_{n,m_2}(x_2) \; \cdots \; A_{n,m_r}(x_r)) = \prod_{1 \leq j < k \leq r} (x_k - x_j)^{m_j m_k},$$

with the $n \times n$ matrix on the left-hand side constructed by concatenation of the r matrices.

Clearly, in the special case $m_1 = \cdots = m_r = 1$ and $n = r$ the evaluation in Lemma 2.6 reduces to the classical Vandermonde determinant.

Lemma 2.7 (Hadamard's inequality) *For the determinant* $\Delta = \det_{1 \leq j,k \leq n}(a_{jk})$ *of an $n \times n$ matrix with complex entries, the following inequality is valid:*

$$|\Delta|^2 \leq \prod_{k=1}^{n}\left(\sum_{j=1}^{n}|a_{jk}|^2\right).$$

Proof In the real case, this follows from the fact that the absolute value of the determinant of real vectors is equal to the volume of the parallelepiped spanned by those vectors. □

The norm of the k-column $(a_{1k}(x), \ldots, a_{nk}(x))^t$ of the matrix $A_{n,m}(x)$ in Lemma 2.6 admits the following estimate:

$$\sum_{j=1}^{n}|a_{jk}(x)|^2 \leq \sum_{\ell=k-1}^{n-1}\binom{\ell}{k-1}^2|x|^{2(\ell-k+1)} \leq (\max\{1,|x|\})^{2n}\sum_{\ell=k-1}^{n-1}\binom{\ell}{k-1}^2$$

$$\leq n^{2k-1}(\max\{1,|x|\})^{2n}.$$

Therefore, for the determinant in the lemma, Hadamard's inequality implies

$$\left|\det(A_{n,m_1}(x_1)\ A_{n,m_2}(x_2)\ \cdots\ A_{n,m_r}(x_r))\right|^2 \leq \prod_{j=1}^{r}n^{m_j^2}(\max\{1,|x_j|\})^{2m_jn}. \quad (2.3)$$

Exercise 2.5 (Mahler [131]) For the discriminant

$$D(P) = a_d^{2d-2}\prod_{1 \leq j < k \leq d}(\alpha_k - \alpha_j)^2 \quad (2.4)$$

of a polynomial $P(x) = a_d\prod_{j=1}^{d}(x - \alpha_j) \in \mathbb{C}[x]$, show the following inequality:

$$|D(P)| \leq d^d\,\mathrm{M}(P)^{2d-2}.$$

Hint Apply Hadamard's inequality to the ordinary Vandermonde determinant

$$\det_{1 \leq j,k \leq d}(\alpha_j^{k-1}) = \prod_{1 \leq j < k \leq d}(\alpha_k - \alpha_j). \qquad □$$

Lemma 2.8 *Suppose that α is an algebraic integer of degree d such that $\alpha^q = \beta^q$ for its algebraic conjugate β and some integer $q > 0$. Then there is an algebraic integer $\tilde{\alpha}$ of smaller degree $\tilde{d}$ for which $\mathrm{M}(\tilde{\alpha}) = \mathrm{M}(\alpha)$.*

Proof Decompose the set of conjugates $\alpha_1, \alpha_2, \ldots, \alpha_d$ of $\alpha = \alpha_1$ into the (disjoint!) union of subsets $A_1, \ldots, A_{\tilde{d}}$ such that $\alpha_j, \alpha_\ell \in A_k$ if and only if $\alpha_j^q = \alpha_\ell^q$. The sets have the same cardinality, which is at least 2 by the hypothesis, hence the total number $\tilde{d}$ of subsets is a proper divisor of d. Another property of any two elements $\alpha_j, \alpha_\ell \in A_k$ is $|\alpha_j| = |\alpha_\ell|$, which clearly follows from comparison of the absolute values in $\alpha_j^q = \alpha_\ell^q$. Now set $\tilde{\alpha}_k = \prod_{j \in A_k}\alpha_j$ for

$k = 1, \ldots, \tilde{d}$. These numbers are clearly algebraic conjugates, and $|\tilde{\alpha}_k| > 1$ is equivalent to $|\alpha_j| > 1$ for each $\alpha_j \in A_k$. This implies that the algebraic integer $\tilde{\alpha}_1$ has degree $\tilde{d} < d$ and Mahler measure $M(\tilde{\alpha}_1) = M(\alpha_1)$. □

Exercise 2.6 Show that if all conjugates of a real algebraic number $\alpha > 1$ have absolute value less than or equal to 1 (in other words, α is a Salem number), then all powers α^q for $q = 1, 2, \ldots$ share the same degree.

One of the crucial ingredients in the proof of Theorem 2.5 is the following arithmetic observation.

Lemma 2.9 (Dobrowolski) *For a monic irreducible non-cyclotomic polynomial* $P(x) = \prod_{j=1}^{d}(x - \alpha_j) \in \mathbb{Z}[x]$ *and a prime* p, *the integer*

$$\prod_{j=1}^{d} P(\alpha_j^p) = \prod_{j=1}^{d} \prod_{k=1}^{d} (\alpha_j^p - \alpha_k)$$

is non-zero and divisible by p^d.

Proof The vanishing of the number would imply that the root α_1 is a root of unity (see Exercise 1.2). The divisibility is a consequence of Fermat's little theorem: $P(\alpha_j^p) \equiv P(\alpha_j)^p \equiv 0 \bmod p$ for $j = 1, \ldots, d$. □

Proof of Theorem 2.5 We will show estimate (2.2) with $c_0 = 2 - \varepsilon$ for all $d > d_0(\varepsilon)$. We can replace, if necessary, our α in question with $\tilde{\alpha}$ as in Lemma 2.8 of the same Mahler measure but smaller degree and satisfying the property that all powers $\tilde{\alpha}^q$ for $q = 1, 2, \ldots$ have the same degree. Indeed, estimate (2.2) for $\tilde{\alpha}$ in place of α implies (2.2) for α itself, because the right-hand side in the estimate is monotone decreasing in d. In summary, we assume that the minimal polynomial $P(x) = \prod_{j=1}^{d}(x - \alpha_j) \in \mathbb{Z}[x]$ of $\alpha = \alpha_1$ is monic (as otherwise its Mahler measure $M(\alpha) = M(P) \geq 2$), and that no integers $q > 0$ and $j \geq 2$ exist such that $\alpha_1^q = \alpha_j^q$ (since we can then reduce α to $\tilde{\alpha}$ using Lemma 2.8).

For $d > d_0(\varepsilon)$, take $m = \log d$, and then choose $k = \lfloor m/\log m \rfloor$ and $s = \lfloor \frac{1}{2}(m/\log m)^2 \rfloor$. Denote by $p_1 = 2, p_2 = 3, \ldots$, the primes numbered in ascending order. Consider the confluent determinant Δ of Lemma 2.6 constructed on the data

$$(x_1, \ldots, x_r) = (\alpha_1, \ldots, \alpha_d, \alpha_1^{p_1}, \ldots, \alpha_d^{p_1}, \ldots, \alpha_1^{p_s}, \ldots, \alpha_d^{p_s})$$

and

$$(m_1, \ldots, m_r) = (\underbrace{k, \ldots, k}_{d \text{ times}}, \underbrace{1, \ldots, 1, \ldots, 1, \ldots, 1}_{sd \text{ times}}).$$

Then Δ is an integer, while estimate (2.3) results in

$$\Delta^2 \leq \prod_{j=1}^{d} n^{k^2+s}(\max\{1, |\alpha_j|\})^{2(k+p_1+\cdots+p_s)n} = n^{d(k^2+s)} \, \mathrm{M}(\alpha)^{2(k+p_1+\cdots+p_s)n}, \quad (2.5)$$

where $n = d(k + s)$. On the other hand, the very same determinant $\Delta \in \mathbb{Z}$ written in the closed form from Lemma 2.6 is non-zero (by Exercise 1.2 and the hypothesis on α which excludes the situation discussed in Lemma 2.8) and divisible by the product

$$\prod_{\ell=1}^{s} \prod_{j=1}^{d} P(\alpha_j^{p_\ell})^k. \quad (2.6)$$

It follows then from Lemma 2.9 that Δ is divisible by $\prod_{\ell=1}^{s} p_\ell^{dk}$, hence

$$\Delta^2 \geq (p_1 \cdots p_s)^{2dk}. \quad (2.7)$$

Finally, using the asymptotical formulae

$$\sum_{\ell=1}^{s} p_\ell = \frac{1}{2} s^2 \log s \, (1 + o(1)) = \frac{m^4}{4 \log^3 m} (1 + o(1))$$

and

$$\sum_{\ell=1}^{s} \log p_\ell = s \log s \, (1 + o(1)) = \frac{m^2}{\log m} (1 + o(1))$$

as $m = \log d \to \infty$, and comparing the estimates in (2.5) and (2.7) we conclude that

$$\begin{aligned}
\log \mathrm{M}(\alpha) &\geq \frac{2dk(\log p_1 + \cdots + \log p_s) - d(k^2 + s) \log n}{2(k + p_1 + \cdots + p_s)n} \\
&= \frac{2k(\log p_1 + \cdots + \log p_s) - (k^2 + s)(\log d + \log(k + s))}{2(k + p_1 + \cdots + p_s)(k + s)} \\
&= \frac{\dfrac{2m^3}{\log^2 m}(1 + o(1)) - \dfrac{3}{2}\dfrac{m^3}{\log^2 m}(1 + o(1))}{\dfrac{m^4}{2\log^3 m}\dfrac{m^2}{2\log^2 m}(1 + o(1))} = \frac{2\log^3 m}{m^3}(1 + o(1)).
\end{aligned}$$

This is precisely the estimate in Theorem 2.5 with $c_0 = 2 - \varepsilon$. $\qquad \square$

2.3 A discriminated problem about discriminants

A classical theorem by Minkowski [141] asserts that the absolute value of the discriminant $D(K)$ of a number field K of degree $[K : \mathbb{Q}] = d$ is bounded below by C^d for some absolute constant $C > 1$. The original estimates for C from the geometry of numbers were overridden by those coming from an analytical method developed by Odlyzko in the 1970s; see the review of the topic in [148] which, in spite of its publication a couple of decades ago, represents the state of the art quite accurately. In the opposite direction it is known that the Minkowski–Odlyzko bounds are best possible in the sense that there are explicit constructions of number fields K whose discriminants (in absolute value) are bounded above by C_1^d with $C_1 \approx 913.493$ according to [134]. The lower bounds remain valid for the discriminant $D(P(x))$ of an irreducible monic polynomial $P(x) = \prod_{j=1}^d (x - \alpha_j) \in \mathbb{Z}[x]$ of degree d as defined in (2.4) with $a_d = 1$, because $D(P(x)) = D(K)$ when the number field $K = \mathbb{Q}(\alpha)$ is monogenic (that is, its ring of integers is generated by the powers $1, \alpha, \alpha^2, \ldots, \alpha^{n-1}$ of a zero $\alpha = \alpha_1 \in \mathbb{C}$ of $P(x)$). Less is known about the sharpness of the estimate for $D(P(x))$. In fact, a question in [148], namely, Open Problem 2.5 attributed there to J.-P. Serre 'and others', specifically asks whether there is an infinite sequence of irreducible monic polynomials $P(x) \in \mathbb{Z}[x]$ of increasing degree d for which $\delta(P) = |D(P(x))|^{1/d}$ — the so-called root discriminants — are bounded above. Odlyzko's comment on the problem in [148] is that 'nothing is known on this topic'. Surprisingly enough, there seems to have been no progress made on this natural question since 1990.

We point out a potential interest in the Odlyzko–Serre problem restricted to certain subclasses of irreducible monic polynomials $P(x) \in \mathbb{Z}[x]$ of degree d. Namely, the knowledge of sharp lower bounds for reciprocal polynomials (that is, for those satisfying $x^d P(1/x) = P(x)$) or for totally real polynomials has potential for the resolution of Lehmer's problem [122], at least for strengthening Dobrowolski's bound for the Mahler measure. The relevant expectations, heuristics and details are outlined in this section.

A few years after Dobrowolski's work [75], Matveev [135] showed that, under some mild conditions on the irreducible polynomial $P(x)$, one can replace the degree d on the right-hand side of the bound (2.2) with the smaller quantity $v = v(P(x)) = d/\delta(P)$:

$$\mathrm{m}(P(x)) \geq \tilde{c}_0 \left(\frac{\log \log v}{\log v} \right)^3. \tag{2.8}$$

Thus, having the root discriminant $\delta(P) = |D(P)|^{1/d}$ being 'sufficiently close' to d improves Dobrowolski's estimate, if not establishes the existence of least Mahler measure. Here we display the following variant of Matveev's observation.

Proposition 2.10 *Assume the following* uniform *estimate for the root discrimi-*

nants of monic reciprocal irreducible non-cyclotomic polynomials $P(x) \in \mathbb{Z}[x]$ of degree d:

$$\delta(P) \geq \frac{d}{\lambda(d)} \tag{2.9}$$

for a non-decreasing function $\lambda(d)$ as $d \to \infty$.

Then for any algebraic number α, which is not a root of unity,

$$\mathrm{m}(\alpha) \geq c_0 \left(\frac{\log \log \lambda(d)}{\log \lambda(d)} \right)^3 \tag{2.10}$$

with some absolute constant $c_0 > 0$, where d is the degree of α.

The choice of constant c_0 is essentially the same as in Theorem 2.5; in particular, it can be set to $\frac{9}{4} - \varepsilon$ following the scheme in [127, 176].

Proof If α is subject to the hypothesis of Lemma 2.8 then we reduce it to a related $\tilde{\alpha}$ with $\mathrm{M}(\tilde{\alpha}) = \mathrm{M}(\alpha)$, smaller degree and the property that all its powers have the same degree as $\tilde{\alpha}$ itself. Then estimate (2.10) for $\tilde{\alpha}$ implies it for α as well, because $\lambda(d)$ is non-decreasing function.

We repeat the argument in our proof of Theorem 2.5 in Section 2.2, except that we set $m = \log \lambda(d)$ for $d > d_0(\varepsilon)$, and observe that the non-zero integer Δ^2 is divisible by the square of the product (2.6) *and* by the product of discriminants

$$D(\alpha)^{k^2} D(\alpha^{p_1}) \cdots D(\alpha^{p_s}).$$

Noticing that $D(\alpha^p)$ is divisible by $D(\alpha) = \delta(P)^d$ for any prime p, or applying estimate (2.9) directly to each factor, we can replace inequality (2.7) with

$$\Delta^2 \geq (p_1 \cdots p_s)^{2dk} \left(\frac{d}{\lambda(d)} \right)^{d(k^2+s)}. \tag{2.11}$$

Now comparison of (2.5) and (2.11) brings us to the estimate

$$\log \mathrm{M}(\alpha) \geq \frac{2k(\log p_1 + \cdots + \log p_s) - (k^2 + s)(\log \lambda(d) + \log(k + s))}{2(k + p_1 + \cdots + p_s)(k + s)},$$

which is of the desired form thanks to the earlier asymptotics computation. $\square$

Proposition 2.10 clearly motivates the following question on discriminants of reciprocal polynomials.

Question 2.11 For an irreducible monic reciprocal polynomial $P(x) \in \mathbb{Z}[x]$ of degree d, how small can its root discriminant be?

The example $P(x) = \Phi_m(x)$, where $m = p_1 p_2 \cdots p_s$ is the product of the first

s primes, due to Scholz [182], complemented by the asymptotics from [193] (see Exercise 2.9 below),

$$\delta(\Phi_m) \sim e^{2\gamma}\frac{d\log\log d}{\log d} \quad \text{as } s \to \infty,$$

where $d = \varphi(m) = \prod_{j=1}^s(p_j - 1)$ and γ is Euler's constant, gives us an upper bound (best known!) for the smallness in Question 2.11. If

$$\delta(P) \geq C_0\frac{d\log\log d}{\log d}$$

with some absolute constant $C_0 > 0$ held true for all irreducible monic reciprocal polynomials $P(x) \in \mathbb{Z}[x]$, then Proposition 2.10 would lead to the improvement

$$m(P(x)) \geq c_0\left(\frac{\log\log\log d}{\log\log d}\right)^3$$

of Dobrowolski's bound. Also notice some evidence of a weaker lower estimate $\delta(P) \geq C\sqrt{d}$ highlighted in [193] in the case of *reducible* polynomials.

It is natural to ask the same question under the extra condition that $P(x)$ is non-cyclotomic, though the existing analysis around Lehmer's problem itself already demonstrates that filtering cyclotomic polynomials out is a tough task.

Question 2.12 For an irreducible monic reciprocal non-cyclotomic polynomial $P(x) \in \mathbb{Z}[x]$ of degree d, how small can its root discriminant be?

Here we may be even more optimistic and offer $|D(P)| \geq d!$ (for almost all such $P(x)$) as a plausible answer, already sufficient for resolution of Lehmer's problem with the help of Proposition 2.10. The expectation can be compared with the upper estimate for $|D(P)|$ in Exercise 2.5 for irreducible $P(x) \in \mathbb{Z}[x]$ of degree d, coming from Hadamard's inequality. The latter inequality (and the sharpness of Hadamard's inequality [73]) suggest that $d\log d$ can be a typical order of magnitude for the bound of $\log|D(P(x))|$ from below. An inspection of the top 100 smallest Mahler measures from [144] indicates that for each polynomial $P(x)$ on the list,

$$0.84840135135 \leq \frac{\log|D(P(x))|}{d\log d} < 1$$

(and the minimum is attained for a polynomial of degree 18), and we always have $D(P(x)) > d!$.

There is a version of Lehmer's question about the existence of a least Salem number (see Exercise 2.6)—an algebraic integer $\alpha > 1$ all of whose conjugates are located on or inside the unit disc. The expectation is that the Mahler measure of Lehmer's polynomial in (1.2) is such a minimal Salem number; on the other hand, polynomials of reasonably small Mahler measure sometimes have several non-real zeros outside the unit disc — see [144].

When all the zeros of an irreducible monic reciprocal polynomial $P(x) \in \mathbb{Z}[x]$ under consideration are on the unit circle or real (a Salem-type case) then the three elementary statements in Exercise 2.7 below reduce the problem of estimating $D(P(x))$ to that for totally real polynomials.

Exercise 2.7 Verify the following claims.

(a) Suppose that $P(x)$ is an irreducible monic reciprocal polynomial with integral coefficients of degree $2d$ and denote by $\alpha_1, \alpha_1^{-1}, \dots, \alpha_d, \alpha_d^{-1}$ its zeros. Then $Q(x) = \prod_{j=1}^{d}(x-(\alpha_j+\alpha_j^{-1}))$ is an irreducible (and monic) polynomial in $\mathbb{Z}[x]$.

(b) Suppose that $Q(x) = \prod_{j=1}^{d}(x - \beta_j) \in \mathbb{Z}[x]$ is irreducible of degree $d \geq 1$. Define α_j to be one of the zeros of $x^2 - \beta_j x + 1$, so that $\alpha_j + \alpha_j^{-1} = \beta_j$, for $j = 1, \dots, d$. Then $P(x) = \prod_{j=1}^{d}(x - \alpha_j)(x - \alpha_j^{-1})$ is an irreducible polynomial in $\mathbb{Z}[x]$.

(c) If irreducible polynomials $P(x) = \prod_{j=1}^{d}(x-\alpha_j)(x-\alpha_j^{-1}) \in \mathbb{Z}[x]$ and $Q(x) = \prod_{j=1}^{d}(x - (\alpha_j + \alpha_j^{-1})) \in \mathbb{Z}[x]$ are as in parts (a) and (b), then $D(Q)^2$ divides $D(P)$.

Hint (c) This follows from

$$((x + 1/x) - (y + 1/y))^2 = (x - y)(1/x - 1/y)(x - 1/y)(1/x - y). \qquad \square$$

The rationale behind reducing the estimation problem to the totally real case is that the bounds for totally real fields (not polynomials!) are somewhat larger than for general number fields. If we denote by $\delta_1(n)$ the minimal root discriminant over totally real fields of degree n, then it is known [134] that $\delta_1^* = \liminf_{n\to\infty} \delta_1(n) < 913.5$. In the opposite direction, this can be compared with the unconditional bound $\delta_1^* \geq 60.8$ from Odlyzko's work and also $\delta_1^* \geq 215.3$ assuming the generalised Riemann hypothesis.

The paper [214] lists all totally real fields of discriminant $< 14^9$; each field appears to be generated by (the powers of) a zero of a monic polynomial. The corresponding reciprocal polynomials all fit the expected estimate $D(P) > d!$ well. To illustrate this, consider the nonic totally real fields of discriminant $< 14^9$ from [214, Table 5]. The sixth polynomial from the table,

$$Q_6(x) = x^9 - x^8 - 8x^7 + 7x^6 + 21x^5 - 15x^4 - 20x^3 + 10x^2 + 5x - 1,$$

corresponds to the cyclotomic polynomial $P(x) = \Phi_{38}(x)$, while the fifth and ninth polynomials,

$$Q_5(x) = x^9 - 2x^8 - 7x^7 + 11x^6 + 18x^5 - 17x^4 - 19x^3 + 6x^2 + 7x + 1,$$

$$Q_9(x) = x^9 - 3x^8 - 4x^7 + 15x^6 + 4x^5 - 22x^4 - x^3 + 10x^2 - 1,$$

correspond to the reciprocal polynomials of Mahler measure $1.91153\ldots$ and $1.84027\ldots$ (both are Salem numbers). The other eight polynomials correspond to the polynomials $P(x)$ of Mahler measure greater than 2.

Chapter notes

Lehmer's problem is settled by Amoroso and Dvornicich [5] in the case when the extension $\mathbb{Q}(\alpha)$ over $\mathbb{Q}$ is abelian.

Another partial resolution of Lehmer's problem is discussed in [80]. There it is established that

$$m(P) > c\left(1 - \frac{1}{d+1}\right)$$

for $c = 0.41623072\ldots$ when P has *odd coefficients*, degree d and no cyclotomic factors, thereby improving the bound of $c = \frac{1}{4}\log 5$ obtained in [35].

In [215], Voutier gives the bound

$$m(\alpha) > \frac{1}{4}\left(\frac{\log\log d}{\log d}\right)^3,$$

valid for all non-zero algebraic α of degree $d \geq 2$, which are not roots of unity. This remains the record estimate when no restriction on α is assumed. The method of proof in [215] is an elaboration of the method from Section 2.2, additionally taking into account lower bounds for discriminants of number fields to estimate some of the terms in the confluent Vandermonde determinant.

In [79], Dubickas proves that for an algebraic number α, which is not a root of a rational number, of degree d one has the estimate

$$|D(\alpha^n)|^{1/(d_n(d_n-1))} \geq M(\alpha)^{n/d - c_1 d^5 (\log d)(\log n)}$$

with some absolute constant $c_1 > 0$, where d_n denotes the degree of α^n. In view of $2 \leq d_n \leq d$ and Dobrowolski's estimate, he obtains as a corollary that $|D(\alpha^n)| \to \infty$ as $n \to \infty$ for any such α.

A conjecture closely related to Lehmer's problem was posed by Schinzel and Zassenhaus [177] in 1965. It states that for the house of an algebraic integer α of degree d, which is not a root of unity, the estimate $\lceil\alpha\rceil \geq 1 + c/d$ holds with some absolute constant $c > 0$. It follows from Exercise 1.12(a) that $\lceil\alpha\rceil \geq 1 + m(\alpha)/d$, so that indeed any bound $m(\alpha) \geq c > 0$ implies the Schinzel–Zassenhaus conjecture. Quite recently, the conjecture was shown to be true by

Dimitrov [71] whose proof makes use of sharp inequalities for the transfinite diameter of 'hedgehog' configuration and a rationality criterion of Bertrandias, but does not appeal to the Mahler measure.

Finally, we would like to point out the online resource [144] developed by Mossinghoff. It not only contains records of small Mahler measures but also outlines of techniques used in their search, as well as numerous useful references for Lehmer's problem and Mahler measures.

Additional exercises

Exercise 2.8 ([157, Section 13]) For the discriminant of a cyclotomic polynomial prove the formula

$$D(\Phi_n(x)) = (-1)^{\varphi(n)/2} n^{\varphi(n)} \prod_{p|n} p^{-\varphi(n)/(p-1)}.$$

Exercise 2.9 ([193]) Show that for the cyclotomic polynomial $\Phi_m(x)$ of index $m = p_1 p_2 \cdots p_s$ being the product of the first s prime numbers,

$$|D(\Phi_m(x))|^{1/d} \sim e^{2\gamma} \frac{d \log \log d}{\log d} \quad \text{as} \quad s \to \infty,$$

where $d = \varphi(m) = \prod_{j=1}^{s}(p_j - 1)$ is the degree of Φ_m and γ is Euler's constant.

Lemma 2.9 demonstrates that the resultant of the minimal polynomials of algebraic numbers α and α^p, which are not roots of unity, is divisible by a high power of p. In general, the resultant of two polynomials $P(x) = a_d \times \prod_{j=1}^{d}(x - \alpha_j) \in \mathbb{C}[x]$ and $Q(x) = b_m \prod_{k=1}^{m}(x - \beta_k) \in \mathbb{C}[x]$ is defined as

$$R(P, Q) = a_d^m b_m^d \prod_{j=1}^{d} \prod_{k=1}^{m} (\alpha_j - \beta_k) = a_d^m \prod_{j=1}^{d} Q(\alpha_j) = (-1)^{dm} b_m^d \prod_{k=1}^{m} P(\beta_k).$$

Exercise 2.10 (a) Verify that $R(Q, P) = (-1)^{\deg P \times \deg Q} R(P, Q)$.
(b) Show that if $P(x) = A(x)Q(x) + B(x)$ and $Q(x) = b_m x^m + \cdots$, then $R(P, Q) = b_m^{\deg P - \deg B} R(B, Q)$.
(c) Show that for three polynomials $P, Q, S \in \mathbb{C}[x]$ the following identity is true:

$$R(P, QS) = R(P, Q)R(P, S).$$

Exercise 2.11 (a) Show that the discriminant of polynomial $P(x) = a_d x^d + \cdots + a_1 x + a_0 \in \mathbb{C}[x]$ can be computed via the formula

$$a_d D(P) = (-1)^{d(d-1)/2} R(P, P'),$$

where $P'(x)$ is the derivative of $P(x)$ and R is the resultant.

(b) Prove the following formula for the discriminant of the product of two polynomials:

$$D(PQ) = D(P)D(Q)R(P, Q)^2.$$

Exercise 2.12 ([157, Section 13]) For the resultant of two cyclotomic polynomials $\Phi_n(x)$ and $\Phi_m(x)$ with $n \leq m$ show that

$$R(\Phi_n, \Phi_m) = \begin{cases} 0 & \text{if } m = n, \\ p^{\varphi(n)} & \text{if } m = np^k \text{ for some } k \geq 1, \\ 1 & \text{otherwise.} \end{cases}$$

3

Multivariate setting

In this chapter, we follow Mahler's steps in defining the measure for polynomials in several variables [130]. We give some examples of multivariate Mahler measures to provide compelling evidence of their 'originality', and we also indicate some intriguing connections to univariate Mahler measures and Lehmer's problem.

3.1 Mahler's invention

There is a natural way to generalise the Mahler measure (1.3) to polynomials in several variables based on the integral representation (1.4). For non-zero $P(x_1, \ldots, x_k) \in \mathbb{C}[x_1, \ldots, x_k]$, following Mahler [130] define

$$M(P(x_1, \ldots, x_k)) = \exp(m(P(x_1, \ldots, x_k))),$$

where the logarithmic Mahler measure

$$\begin{aligned}
m(P(x_1, \ldots, x_k)) &= \int \cdots \int_{[0,1]^k} \log |P(e^{2\pi i t_1}, \ldots, e^{2\pi i t_k})| \, dt_1 \cdots dt_k \\
&= \frac{1}{(2\pi i)^k} \int \cdots \int_{|x_1| = \cdots = |x_k| = 1} \log |P(x_1, \ldots, x_k)| \, \frac{dx_1}{x_1} \cdots \frac{dx_k}{x_k}
\end{aligned}$$
(3.1)

is the arithmetic average of $\log |P(x_1, \ldots, x_k)|$ over the k-dimensional torus $\mathbb{T}^k : |x_1| = \cdots = |x_k| = 1$. The Mahler measure of the zero polynomial is set to be 0. The definition works well more generally for Laurent polynomials $P(x_1, \ldots, x_k) \in \mathbb{C}[x_1^{\pm 1}, \ldots, x_k^{\pm 1}]$ and, in view of the property $M(PQ) = M(P) M(Q)$, can be further extended to rational functions.

Since the polynomial P may vanish on the torus $\mathbb{T}^k$, the integrand in (3.1) can have singularities. The following proposition shows that this integral in fact always converges.

Proposition 3.1 *The integral* (3.1) *defining* $\mathrm{m}(P)$ *converges for every non-zero Laurent polynomial* $P \in \mathbb{C}[x_1^{\pm 1}, \ldots, x_k^{\pm 1}]$.

We first prove Proposition 3.1 in a special case, when the polynomial is (essentially) monic with respect to one of its variables x_j. The full proof of Proposition 3.1 is postponed to Section 3.2, after we introduce the Newton polytope associated to a polynomial.

Let $P(x_1, \ldots, x_k) \in \mathbb{C}[x_1^{\pm 1}, \ldots, x_k^{\pm 1}]$ be a non-zero Laurent polynomial. Write $a(x_1, \ldots, x_{k-1})$ for the leading coefficient of P viewed as a polynomial in x_k. We say that P is quasi-monic with respect to x_k if $a(x_1, \ldots, x_{k-1}) = c x_1^{e_1} \cdots x_{k-1}^{e_{k-1}}$ for some $c \in \mathbb{C}^{\times}$ and $e_1, \ldots, e_{k-1} \in \mathbb{Z}$.

Proof of Proposition 3.1 in the quasi-monic case   Let $P(x_1, \ldots, x_k)$ be quasi-monic with respect to some variable x_j. Re-indexing the variables, we may assume $j = k$. Multiplying P by a monomial does not change the value of $\log |P|$ on the torus, so we may assume that P is a true (Laurent) polynomial in x_k and that $a(x_1, \ldots, x_{k-1}) = c \in \mathbb{C}^{\times}$. Then write

$$P(x_1, x_2, \ldots, x_k) = c \prod_{j=1}^{d} (x_k - \alpha_j(x_1, \ldots, x_{k-1})), \qquad (3.2)$$

where the α_j denote the roots of P seen as a polynomial in x_k. Substituting this product into (3.1) and performing the integration with respect to x_k using Jensen's formula, we have

$$\mathrm{m}(P) = \log |c| + \int \cdots \int_{[0,1]^{k-1}} A(e^{2\pi i t_1}, \ldots, e^{2\pi i t_{k-1}}) \, dt_1 \cdots dt_{k-1},$$

where

$$A(x_1, \ldots, x_{k-1}) = \sum_{j=1}^{d} \log^+ |\alpha_j(x_1, \ldots, x_{k-1})|.$$

Since the zeros of a polynomial depend continuously on its coefficients, the function A is continuous on the torus $\mathbb{T}^{k-1}$. This shows the convergence of the integral defining $\mathrm{m}(P)$. □

Exercise 3.1 (a) Show that for any Laurent polynomial $P(x_1, \ldots, x_k)$,

$$\mathrm{m}(P(x_1, \ldots, x_k)) = \mathrm{m}(P(x_1^{-1}, \ldots, x_k^{-1})).$$

(b) Verify that in the case of a polynomial P with *real* coefficients, the intersection of the zero locus $P(x_1, \ldots, x_k) = 0$ with the torus $\mathbb{T}^k$ coincides with the intersection of $P(x_1^{-1}, \ldots, x_k^{-1}) = 0$ with $\mathbb{T}^k$. In other words, the intersection is contained in the set of solutions of the system

$$P(x_1, \ldots, x_k) = P(x_1^{-1}, \ldots, x_k^{-1}) = 0.$$

We now discuss estimates for $m(P)$ in terms of the coefficients of the polynomial. In the next statements we deal with polynomials

$$P(x_1, \ldots, x_k) = \sum_{\substack{n_j=0 \\ j=1,\ldots,k}}^{d_j} a_{n_1,\ldots,n_k} x_1^{n_1} \cdots x_k^{n_k}$$

and attach to them the standard characteristics of height $H(P)$ and length $L(P)$ as follows:

$$H(P) = \max_{\substack{0 \le n_j \le d_j \\ j=1,\ldots,k}} \{|a_{n_1,\ldots,n_k}|\} \quad \text{and} \quad L(P) = \sum_{\substack{n_j=0 \\ j=1,\ldots,k}}^{d_j} |a_{n_1,\ldots,n_k}|.$$

Proposition 3.2 (Mahler [130]) *For a polynomial $P \in \mathbb{C}[x_1, \ldots, x_k]$, we have*

$$M(P) \le L(P) \le 2^{d_1 + \cdots + d_k} M(P),$$

where $d_1, \ldots, d_k$ are degrees of P with respect to the corresponding variables $x_1, \ldots, x_k$.

Proof Viewing P as a polynomial in variables $x_{j+1}, \ldots, x_k$ with coefficients in $\mathbb{C}[x_1, \ldots, x_j]$, where $j = 1, \ldots, k-1$, and applying Lemma 1.6 repeatedly, we arrive at the estimate

$$|a_{n_1,\ldots,n_k}| \le \binom{d_1}{n_1} \cdots \binom{d_k}{n_k} M(P), \quad \text{where} \ 0 \le n_j \le d_j, \ j = 1, \ldots, k.$$

The estimate for $L(P)$ from above is the result of summing the inequalities over all n_j, where $j = 1, \ldots, k$.

The inequality $M(P) \le L(P)$ follows from integrating the trivial estimate $|P(e^{2\pi i t_1}, \ldots, e^{2\pi i t_k})| \le L(P)$. $\qquad \square$

Exercise 3.2 (a) Demonstrate that $M(\partial P/\partial x_j) \le d_j M(P)$ for $j = 1, \ldots, k$.

(b) Show that if $v = v(P)$ is the number of variables x_j, for which $\partial P/\partial x_j$ does not vanish identically, then

$$H(P) \le 2^{d_1 + \cdots + d_k - v} M(P).$$

(c) In the other direction, show that

$$M(P) \le ((d_1 + 1) \cdots (d_k + 1))^{1/2} H(P).$$

Hint   (c) Use the classical inequality

$$M(P) \le \left(\int \cdots \int_{[0,1]^k} |P(e^{2\pi i t_1}, \ldots, e^{2\pi i t_k})|^2 \, dt_1 \cdots dt_k \right)^{1/2}$$

followed by Parseval's equation

$$\int \cdots \int_{[0,1]^k} |P(e^{2\pi i t_1}, \ldots, e^{2\pi i t_k})|^2 \, dt_1 \cdots dt_k = \sum_{\substack{n_j=0 \\ j=1,\ldots,k}}^{d_j} |a_{n_1,\ldots,n_k}|^2$$

$$\le (d_1 + 1) \cdots (d_k + 1) H(P)^2. \qquad \square$$

The following is a multidimensional generalisation of Proposition 1.8.

Proposition 3.3 (Mahler [130]) *For polynomials* $P_1, \ldots, P_m \in \mathbb{C}[x_1, \ldots, x_k]$ *of corresponding total degrees* $d_1, \ldots, d_m$, *the following inequalities are true*:

$$L(P_1 \cdots P_m) \le L(P_1) \cdots L(P_m) \le 2^{d_1 + \cdots + d_m} L(P_1 \cdots P_m).$$

Proof   The inequality $L(P_1 \cdots P_m) \le L(P_1) \cdots L(P_m)$ is nearly trivial, while the upper bound follows from repeated application of Proposition 3.2 to each of the polynomials $P_1, \ldots, P_m$. $\square$

3.2 Newton polytopes

We introduce in this section the Newton polytope associated to a multivariate polynomial, which is a fundamental tool in the study of polynomials and Mahler measures. We will then use it to give a full proof of Proposition 3.1.

We work with Laurent polynomials in k variables $x_1, \ldots, x_k$. Given a k-tuple $\boldsymbol{j} = (j_1, \ldots, j_k) \in \mathbb{Z}^k$, we denote by $\boldsymbol{x}^{\boldsymbol{j}} = x_1^{j_1} \cdots x_k^{j_k}$ the corresponding monomial, and we say that $\boldsymbol{j}$ is the exponent of $\boldsymbol{x}^{\boldsymbol{j}}$. By definition, the Newton polytope N_P of a Laurent polynomial $P \in \mathbb{C}[x_1^{\pm 1}, \ldots, x_k^{\pm 1}]$ is the convex hull of the exponents of the monomials appearing in P. It is a convex polytope in $\mathbb{R}^k$ whose vertices (or extreme points) lie in the lattice $\mathbb{Z}^k$. In the case of two-variable polynomials, the Newton polytope is traditionally called the Newton polygon. Some examples for trinomials are given in Figure 3.1.

The Newton polygon of a two-variable polynomial $P(x, y)$ provides one with interesting information on the algebraic curve $P(x, y) = 0$. By a classical result of Baker [10] (see also [76, Section 2]), if P is irreducible, then the genus g of the curve $P(x, y) = 0$ does not exceed the number h of interior lattice points of N_P. Moreover, under the generic condition that every side polynomial of P has distinct roots, we have $g = h$ (see [165, Section 8] for the definition of

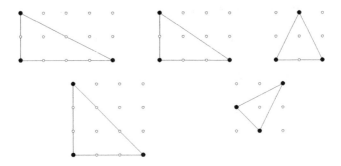

Figure 3.1 Newton polygons for polynomials $P(x, y) = a + bx^4 + cy^2$, $a + bx^3 + cy^2$, $a + bx^2 + cxy^2$, $a + bx^3 + cy^3$ and $ax + by + cx^2y^2$

side polynomials). For example, all curves whose polygons are represented in Figure 3.1 have genus 1 for a generic set of coefficients a, b, c.

Further, if C_P denotes the zero locus of P in $(\mathbb{C}^\times)^2$, then the equality $g = h$ holds if and only if the Zariski closure of C_P in a certain toric surface is non-singular (see [12, Theorem 4] for the precise statement). This toric surface is a compactification of $(\mathbb{C}^\times)^2$ depending only on the Newton polygon N_P, and not on the polynomial itself. For example, if N_P is a triangle with vertices $(0, 0)$, $(d, 0)$ and $(0, d)$, then the toric surface is the projective plane $\mathbb{P}^2(\mathbb{C})$, and we recover the classical fact that a projective plane curve of degree d has genus at most $(d - 1)(d - 2)/2$, with equality if and only if the curve is non-singular. If N_P is a rectangle with vertices $(0, 0)$, $(d, 0)$, $(0, e)$ and (d, e), then the toric surface is $\mathbb{P}^1(\mathbb{C}) \times \mathbb{P}^1(\mathbb{C})$ and the generic genus is $(d - 1)(e - 1)$. The reader may consult [88] for more information on toric varieties.

If $P(x_1, \ldots, x_k) = \sum_{j \in \mathbb{Z}^k} a_j \boldsymbol{x}^j$ is a Laurent polynomial and $g \colon \mathbb{Z}^k \to \mathbb{Z}^k$ is a bijective affine transformation, define $gP = \sum_{j \in \mathbb{Z}^k} a_j \boldsymbol{x}^{gj}$. From the definition, we see that $N_{gP} = g(N_P)$. At the same time the group $\mathrm{GL}_k(\mathbb{Z})$ acts naturally on the torus $\mathbb{T}^k \colon |x_1| = \cdots = |x_k| = 1$ by coordinate transformations,

$$g = (g_{j\ell})_{1 \leq j, \ell \leq k} \in \mathrm{GL}_k(\mathbb{Z}) \colon \begin{pmatrix} x_1 \\ x_2 \\ \vdots \\ x_k \end{pmatrix} \mapsto \begin{pmatrix} x_1^{g_{11}} x_2^{g_{12}} \cdots x_k^{g_{1k}} \\ x_1^{g_{21}} x_2^{g_{22}} \cdots x_k^{g_{2k}} \\ \vdots \\ x_1^{g_{k1}} x_2^{g_{k2}} \cdots x_k^{g_{kk}} \end{pmatrix}.$$

Note that the two actions are compatible in the sense that $(gP)(\boldsymbol{x}) = P({}^t g \boldsymbol{x})$ for every g in $\mathrm{GL}_k(\mathbb{Z})$. The corresponding change of variables in the integral (3.1) shows that if the integral defining $\mathrm{m}(gP)$ exists for some $g \colon \mathbb{Z}^k \to \mathbb{Z}^k$ as above, then it exists for m(P) as well, and the two integrals are equal. (In fact, the latter property is true even for the $GL_k(\mathbb{Q})$ action: the Mahler measure remains unchanged under $x_j \mapsto x_j^a$ for some j and any non-zero $a \in \mathbb{Z}$.)

Proof of Proposition 3.1   Let $P(x_1, \ldots, x_k) \neq 0$ be an arbitrary Laurent polynomial in $x_1, \ldots, x_k$. Let $v \in \mathbb{Z}^k$ be any vertex of the Newton polytope N_P (see Figure 3.2 for an example in the case $k = 2$). Choose any rational hyperplane H in $\mathbb{Q}^k$ such that $v + H$ intersects N_P only at v. Choose a $\mathbb{Z}$-basis $(h_1, \ldots, h_{k-1})$ of $H_{\mathbb{Z}} = H \cap \mathbb{Z}^k$. The abelian group $\mathbb{Z}^k / H_{\mathbb{Z}}$ is torsion-free and has rank 1, hence it is isomorphic to $\mathbb{Z}$; this means that there exists a vector h_k such that $(h_1, \ldots, h_k)$ form a basis of $\mathbb{Z}^k$. We choose h_k pointing *outside* N_P.

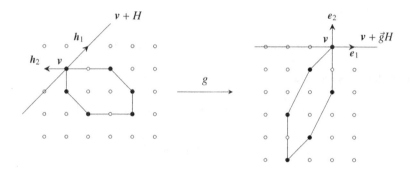

Figure 3.2 Construction of the affine transformation g

Now, let $\vec{g} \colon \mathbb{Z}^k \to \mathbb{Z}^k$ be the linear map sending $(h_1, \ldots, h_k)$ to the canonical basis $(e_1, \ldots, e_k)$ of $\mathbb{Z}^k$. Moreover, let $g \colon \mathbb{Z}^k \to \mathbb{Z}^k$ be the affine transformation that fixes the vertex v and is induced by $\vec{g}$. By construction, the polynomial gP is quasi-monic with respect to x_k. By what we proved in Section 3.1, the integral defining m(gP) converges. The discussion above shows that m(P) also converges, as required. □

Our proof of Proposition 3.1 automatically shows that the Mahler measure of a multivariate polynomial P depends *continuously* on the coefficients of P, provided that the Newton polytope of P does not change. In fact, continuity also holds if we only require the Newton polytope to be contained in a fixed box $[-d, d]^k$ (see [40]).

Exercise 3.3 Show that m(P) ≥ 0 for a polynomial $P(x_1, \ldots, x_k)$ with *integer* coefficients.

3.3 Special Mahler measures

The main difficulty in computing the Mahler measure of multivariate polynomials is the lack of Jensen's formula (though the latter gives an efficient reduction to integrals over $(k-1)$-dimensional domains). Historically, the first computation of such Mahler measures was performed by Smyth [38, Appendix 1] and we reproduce his examples in Proposition 3.4 and Exercise 3.5 below.

Proposition 3.4 (Smyth (1981)) *We have*

$$m(1 + x + y) = \frac{3\sqrt{3}}{4\pi} L(\chi_{-3}, 2),$$

where $\chi_{-3}(n) = \left(\frac{-3}{n}\right)$ is the quadratic character modulo 3.

Proof Notice that

$$\int_0^1 \log|1 + x + e^{2\pi i s}| \, ds = \log^+|1 + x|$$

from Jensen's formula, so that

$$
\begin{aligned}
m(1 + x + y) &= \int_{-1/2}^{1/2} \log^+|1 + e^{2\pi i t}| \, dt = \int_{-1/3}^{1/3} \log|1 + e^{2\pi i t}| \, dt \\
&= \mathrm{Re} \int_{-1/3}^{1/3} \log(1 + e^{2\pi i t}) \, dt = \mathrm{Re} \int_{-1/3}^{1/3} \sum_{n=1}^{\infty} \frac{(-1)^{n-1} e^{2\pi i n t}}{n} \, dt \\
&= \sum_{n=1}^{\infty} \frac{(-1)^{n-1} e^{2\pi i n t}}{2\pi i n^2} \Big|_{t=-1/3}^{t=1/3} = \frac{1}{\pi} \sum_{n=1}^{\infty} \frac{(-1)^{n-1} \sin\frac{2\pi n}{3}}{n^2}.
\end{aligned}
$$

It remains to use $\sin\frac{2\pi n}{3} = \frac{\sqrt{3}}{2}\chi_{-3}(n)$ and the multiplicativity of the character χ_{-3}. □

Exercise 3.4 ([39]) Show that

$$m(1 + x + y - xy) = \frac{2}{\pi} L(\chi_{-4}, 2),$$

where $\chi_{-4}(n) = \left(\frac{-4}{n}\right)$ is the quadratic character modulo 4, so that

$$L(\chi_{-4}, 2) = \sum_{n=0}^{\infty} \frac{(-1)^n}{(2n+1)^2} \tag{3.3}$$

is Catalan's constant.

Exercise 3.5 (Smyth (1981)) Show that

$$m(1 + x + y + z) = \frac{7}{2\pi^2} \zeta(3).$$

Hint The simple change of variable $(x, y, z) \mapsto (x, yz, z)$ in the integral defining the Mahler measure leads to

$$m(1 + x + y + z) = m(1 + x + (1 + y)z),$$

and the latter is equal to $m(\max\{|1 + x|, |1 + y|\})$ because

$$m(a + bz) = \log|a| + \max\{0, \log|b/a|\} = \max\{\log|a|, \log|b|\}$$

from Jensen's formula. □

We can also compute the intersection of the zero loci of the polynomials $1 + x + y$ and $1 + x + y + z$ with the corresponding two- and three-dimensional tori, using the result of Exercise 3.1. By solving the system

$$1 + x + y = 1 + x^{-1} + y^{-1} = 0$$

we find that the solutions are $(x, y) = (e^{2\pi i/3}, e^{-2\pi i/3})$ and $(e^{-2\pi i/3}, e^{2\pi i/3})$, so that they all belong to the cyclotomic extension $\mathbb{Q}(e^{2\pi i/3})$ of $\mathbb{Q}$. The solutions to

$$1 + x + y + z = 1 + x^{-1} + y^{-1} + z^{-1} = 0$$

on the torus $|x| = |y| = |z| = 1$ are given by $x = -1$, $y + z = 0$ intersecting with the two-dimensional torus $|y| = |z| = 1$, and two cyclic permutations of the set. The corresponding points and planes are all defined over $\mathbb{Q}$. The two fields of definitions, $\mathbb{Q}(e^{2\pi i/3})$ and $\mathbb{Q}$, respectively, are clearly related to the L-functions $L(\chi_{-3}, s)$ and $\zeta(s)$ featured in the Mahler measure evaluations.

Exercise 3.6 (Wan (2011)) For real $k > 0$, show that

$$m((1 + x)(1 + y) + kz) = \log k + \frac{8}{\pi^2} \int_k^4 \frac{\arccos(k/t) \log(t/(2\sqrt{k}))}{\sqrt{16 - t^2}} \, dt$$

if $0 < k \le 4$, and $m((1 + x)(1 + y) + kz) = \log k$ if $k > 4$.

3.4 Limits of multivariate Mahler measures

In spite of the visible disjointness of values of the logarithmic multivariate Mahler measure in the last section from those of $m(P(x))$, where $P(x) \in \mathbb{Z}[x]$ — the latter are always $\mathbb{Z}$-linear combinations of the logarithms of algebraic numbers — they are connected by an analytic process. The following result, due to Boyd [38] and Lawton [121], clarifies this connection.

Theorem 3.5 *For a polynomial $P(x, y)$ in two variables, the limit*

$$\lim_{m \to \infty} m(P(x, x^m))$$

exists and is equal to $m(P(x, y))$.

We treat here only an easier case of the theorem, under the severe restriction on $P(x, y)$ not to vanish on the torus $|x| = |y| = 1$, advising the careful follower to consult with [38, Appendix 3] for the general situation. In our proof we make use of the following fact.

Lemma 3.6 *For* $P(x, y) \in \mathbb{C}[x, y]$, *the convergence of the regular Riemann sums*

$$\frac{1}{n} \sum_{k \bmod n} \log |P(e^{2\pi i\lambda/n} \zeta_n^k, y)| \to \int_0^1 \log |P(e^{2\pi it}, y)| \, dt \quad \text{as } n \to \infty$$

is uniform in $y : |y| = 1$ *and* $\lambda \in \mathbb{R}$.

Proof Write $P(x, y) = a_d(y) \prod_{j=1}^d (x - \alpha_j(y))$, where $\alpha_j(y)$ are the branches of the algebraic function defined by $P(\alpha(y), y) = 0$. The idea is to apply the limit

$$\frac{1}{n} \sum_{k \bmod n} \log |e^{2\pi i\lambda/n} \zeta_n^k - \alpha(y)| = \frac{\log |e^{2\pi i\lambda} - \alpha(y)^n|}{n}$$

$$\to \log^+ |\alpha(y)| = \max\{0, \log |\alpha(y)|\} \quad \text{as } n \to \infty$$

to the individual terms in the representation of $\log |P(x, y)|$.

Note that

$$1 - |\alpha(y)| \le |e^{2\pi i\lambda} - \alpha(y)^n| \le 2$$

if $|\alpha(y)| \le 1$, and similarly

$$1 - |\alpha(y)|^{-1} \le |e^{2\pi i\lambda} - \alpha(y)^n| - n \log |\alpha(y)| = |e^{-2\pi i\lambda} - (1/\alpha(y))^n| \le 2$$

if $|\alpha(y)| \ge 1$. This implies that the convergence

$$\frac{\log |e^{2\pi i\lambda} - \alpha(y)^n|}{n} \to \log^+ |\alpha(y)| \quad \text{as } n \to \infty$$

is uniform in y from the domain where $\min\{|\alpha(y)|, 1/|\alpha(y)|\} \le 1 - \rho$ with $\rho > 0$ fixed. The existence of such a ρ follows from the non-vanishing of $P(x, y)$ on the (compact!) torus $|x| = |y| = 1$.   □

Proof of Theorem 3.5 under the severe constraint on P For fixed m, compute the Mahler measure of $P(x, x^m)$ using formula (1.5) with $\omega_n = 1$ along the subsequence $mn \to \infty$:

$$
\begin{aligned}
\mathrm{m}(P(x, x^m)) &= \lim_{n \to \infty} \frac{1}{mn} \sum_{k \bmod mn} \log |P(\zeta_{mn}^k, \zeta_{mn}^{mk})| \\
&= \lim_{n \to \infty} \frac{1}{mn} \sum_{a \bmod m} \sum_{b \bmod n} \log |P(\zeta_{mn}^{an+b}, \zeta_{mn}^{mb})| \\
&= \frac{1}{m} \sum_{a \bmod m} \lim_{n \to \infty} \frac{1}{n} \sum_{b \bmod n} \log |P(\zeta_{mn}^b \zeta_m^a, \zeta_n^b)|.
\end{aligned}
$$

Now, for y with $|y| = 1$, the limit

$$
\lim_{m \to \infty} \frac{1}{m} \sum_{a \bmod m} \log |P(\zeta_{mn}^b \zeta_m^a, y)|
$$

exists, does not depend on b and n and is uniform in y by Lemma 3.6 (in which we take $\lambda = b/n$); this limit is equal to

$$
\lim_{m \to \infty} \frac{1}{m} \sum_{a \bmod m} \log |P(\zeta_m^a, y)|.
$$

The repeated limit

$$
\begin{aligned}
&\lim_{n \to \infty} \frac{1}{n} \sum_{b \bmod n} \lim_{m \to \infty} \frac{1}{m} \sum_{a \bmod m} \log |P(\zeta_{mn}^b \zeta_m^a, \zeta_n^b)| \\
&= \lim_{n \to \infty} \frac{1}{n} \sum_{b \bmod n} \lim_{m \to \infty} \frac{1}{m} \sum_{a \bmod m} \log |P(\zeta_m^a, \zeta_n^b)|
\end{aligned}
$$

has an internal limit uniform in $y = \zeta_n^b$, in other words, in n, and the latter limit is the regular Riemann sum for the double integral defining $\mathrm{m}(P(x, y))$:

$$
\mathrm{m}(P(x, y)) = \lim_{\substack{m \to \infty \\ n \to \infty}} \frac{1}{m} \sum_{a \bmod m} \frac{1}{n} \sum_{b \bmod n} \log |P(\zeta_m^a, \zeta_n^b)|.
$$

Therefore, we can interchange the limits in the former representation to arrive at

$$
\begin{aligned}
\lim_{m \to \infty} \mathrm{m}(P(x, x^m)) &= \lim_{m \to \infty} \frac{1}{m} \sum_{a \bmod m} \lim_{n \to \infty} \frac{1}{n} \sum_{b \bmod n} \log |P(\zeta_{mn}^b \zeta_m^a, \zeta_n^b)| \\
&= \lim_{n \to \infty} \frac{1}{n} \sum_{b \bmod n} \lim_{m \to \infty} \frac{1}{m} \sum_{a \bmod m} \log |P(\zeta_{mn}^b \zeta_m^a, \zeta_n^b)| \\
&= \mathrm{m}(P(x, y)),
\end{aligned}
$$

as required. $\qquad \square$

Exercise 3.7 ([38, Appendix 2]) Demonstrate that

$$\mathrm{m}(1 + x + x^m) = \mathrm{m}(1 + x + y) + \frac{c(m)}{m^2} + O(m^{-3}) \quad \text{as } m \to \infty,$$

where

$$c(m) = \frac{\pi}{6\sqrt{3}} \text{ if } m \equiv 0, 1 \bmod 3 \quad \text{and} \quad c(m) = -\frac{\pi}{2\sqrt{3}} \text{ if } m \equiv 2 \bmod 3.$$

The argument of our proof of Theorem 3.5 can be inductively pushed further to establish the following generalisation (cf. [38, Appendix 4]).

Theorem 3.7 *For a polynomial $P(x_1, \ldots, x_k)$ in k variables, the iterated limit*

$$\lim_{m_2 \to \infty} \cdots \lim_{m_k \to \infty} \mathrm{m}(P(x, x^{m_2}, \ldots, x^{m_k}))$$

of univariate Mahler measures exists and is equal to $\mathrm{m}(P(x_1, \ldots, x_k))$.

3.5 Reciprocal vs non-reciprocal polynomials

Theorems 3.5 and 3.7 suggest looking for small multivariate Mahler measures as potential counterexamples to the expected answer to Lehmer's problem. Indeed, if we manage to construct a polynomial $P(x_1, \ldots, x_k)$ whose Mahler measure $M(P)$ is smaller than that in (1.2), then for (any) sufficiently large set of integers $m_2, \ldots, m_k$ the induced single-variable polynomial

$$P_{m_2, \ldots, m_k}(x) = P(x, x^{m_2}, \ldots, x^{m_k}) \tag{3.4}$$

will also have measure smaller than (1.2). Thanks to Smyth's theorem, even in its weaker form in Theorem 2.2, this may only happen if polynomials (3.4) are reciprocal (and non-cyclotomic). It is not hard to see that the reciprocality of these polynomials is equivalent to the property

$$x_1^{d_1} x_2^{d_2} \cdots x_k^{d_k} P(1/x_1, 1/x_2, \ldots, 1/x_k) = P(x_1, x_2, \ldots, x_k) \tag{3.5}$$

for some integral exponents $d_1, d_2, \ldots, d_k$. (Notice that relation (3.5) immediately implies central symmetry of the Newton polygon of P.) Thus, our multivariate polynomials with small Mahler measure (at least, smaller than the constant $x_0 = 1.32471795\ldots$ in Theorem 2.1) must satisfy (3.5). However, we have to impose some further conditions to make $M(P(x_1, \ldots, x_k)) < x_0$ plausible.

Indeed, as seen in Section 3.2, the action of the group $GL_k(\mathbb{Z})$ (and even of $GL_k(\mathbb{Q})$) on the torus $\mathbb{T}^k : |x_1| = \cdots = |x_k| = 1$ by coordinate transformations corresponds to changes of variable in (3.1) and preserves the (logarithmic) Mahler measure of a (Laurent) polynomial. Therefore, we need property

(3.5) to be satisfied for every $\mathrm{GL}_k(\mathbb{Z})$-transformation of $P(x_1, \ldots, x_k)$ in order to be in the running for $\mathrm{M}(P(x_1, \ldots, x_k)) < x_0$. Furthermore, we would like to have all (3.4) coming from all such transformed polynomials to be monic (as otherwise we have $\mathrm{M}(P_{m_2,\ldots,m_k}) \geq 2$ for trivial reasons). This latter property originates from the temperedness of $P(x_1, \ldots, x_k)$ — a property we will discuss later (see the notes for Chapters 7 and 8). Very roughly, it means that the logarithmic Mahler measures of the polynomials attached to all $(k-1)$-dimensional faces of the Newton polytope of P are zero. In the case of two variables, a polynomial $P(x, y)$ is said to be tempered if the single-variable polynomials corresponding to the sides of its Newton polygon are cyclotomic. We investigate such situations in Exercises 3.10, 3.11 below.

For the moment, we restrict our consideration to a particular family of two-variate Laurent polynomials $P_k(x, y) = x + 1/x + y + 1/y + k$ that pass all such tests, and to their logarithmic Mahler measures

$$\mu(k) = \mathrm{m}\left(x + \frac{1}{x} + y + \frac{1}{y} + k\right)$$

for integers k. In fact, the change of variables $(x, y) \mapsto (-x, -y)$ in the corresponding integral (3.1) for $\mu(k)$ shows that $\mu(k) = \mu(|k|)$, so that it is enough to deal with integers $k \geq 0$. By computing the value $\mu(1) = 0.25133043\ldots$ numerically, one finds out that it is indeed less than $\log x_0 = 0.28119957\ldots$. (This and three more 'competitive' examples of two-variate Mahler measures are given by Boyd in [39, p. 43].)

The Mahler measures $\mu(k)$ are seen to be a two-variable extension of what, in Section 1.4, was the family $\lambda^+(k)$. The substitution

$$x = \frac{XY}{1+X}, \quad y = -\frac{X(k+Y)}{1+X} \tag{3.6}$$

brings us to a different family $\tilde{P}_k(X, Y) = Y^2 + kY + X + 1/X + 2$, which is also tempered (whatever it means for the moment) but not reciprocal; in particular, $\tilde{\mu}(k) \geq \log x_0$ for x_0 from Theorem 2.1, where we set

$$\tilde{\mu}(k) = \mathrm{m}\left(Y^2 + kY + X + \frac{1}{X} + 2\right).$$

Because of the invariance of the torus under $(X, Y) \mapsto (X, -Y)$, we also get $\tilde{\mu}(k) = \tilde{\mu}(|k|)$. Up to the $\mathrm{GL}_2(\mathbb{Q})$-action, this non-reciprocal family was suggested in relation to $P_k(x, y)$ in the paper [19].

The substitution (3.6) corresponds to the isomorphism of the corresponding *elliptic curves* $P_k(x, y) = 0$ and $\tilde{P}_k(X, Y) = 0$. Though this isomorphism does not originate from a $\mathrm{GL}_2(\mathbb{Z})$-transformation, it leads to a coincidence of the

Mahler measures. To prove this, we treat both families as depending on the real parameter $k > 0$ continuously.

Proposition 3.8 *For real $k \geq 5$, we have $\mu(k) = \tilde{\mu}(k)$.*

This theorem means that the Mahler measures $\mu(k)$ of a 'strongly' reciprocal family cannot be too small for many values of k as they coincide with non-reciprocal Mahler measures (which typically have a tendency to be larger, as Smyth's theorem suggests). Such coincidences of the Mahler measures corresponding to reciprocal and non-reciprocal multivariate polynomials seem to be a general phenomenon, which closes gates to attacks on Lehmer's problem from the direction of several variables.

Our proof below of Proposition 3.8 compares the derivatives with respect to k of the two Mahler measures in question. As a by-product, we deduce that the derivatives are *periods* of related elliptic curves.

Lemma 3.9 *For real $k > 0$, $k \neq 4$, we have*

$$\frac{d\mu(k)}{dk} = \frac{1}{\pi} \, \mathrm{Re} \int_0^1 \frac{dt}{\sqrt{t(1-t)(k^2 - 16t)}}.$$

Proof We have

$$\mu(k) = \mathrm{m}(y^{-1}(y^2 + (x + x^{-1} + k)y + 1)) = \mathrm{m}(y^2 + b(x)y + 1),$$

where $b(x) = b_k(x) = x + x^{-1} + k$. Notice that $b_k(x)$ is real valued on the circle $|x| = 1$, and $|b_k(x) - k| \leq 2$. The polynomial $y^2 + b(x)y + 1$ has two zeros, $y_1(x) = -(b(x) + \sqrt{\delta(x)})/2$ and $y_2(x) = -(b(x) - \sqrt{\delta(x)})/2$, where $\delta(x) = \delta_k(x) = b(x)^2 - 4$ is real valued on $|x| = 1$. They satisfy $y_1(x)y_2(x) = 1$ by Viète's theorem, so that $|y_1(x)| = |y_2(x)| = 1$ when $\delta(x) \leq 0$ and $|y_1(x)| > 1 > |y_2(x)|$ when $\delta(x) > 0$. In each of these cases we have

$$\max\{1, |y_1(x)|\} = |y_1(x)| \quad \text{and} \quad \max\{1, |y_2(x)|\} = 1.$$

Therefore, applying Jensen's formula and using the symmetry $y_1(x) = y_1(x^{-1})$ we deduce that

$$\mu(k) = \mathrm{m}((y - y_1(x))(y - y_2(x)))$$
$$= \frac{1}{(2\pi i)^2} \iint_{|x|=|y|=1} \log |(y - y_1(x))(y - y_2(x))| \frac{dx}{x} \frac{dy}{y}$$
$$= \frac{1}{2\pi i} \int_{|x|=1} \log^+ |y_1(x)| \frac{dx}{x} + \frac{1}{2\pi i} \int_{|x|=1} \log^+ |y_2(x)| \frac{dx}{x}$$

$$= \frac{1}{2\pi i} \int_{|x|=1} \log |y_1(x)| \frac{dx}{x} = \frac{1}{\pi} \int_0^\pi \log |y_1(e^{i\theta})| \, d\theta$$

$$= \frac{1}{\pi} \, \mathrm{Re} \int_0^\pi \log \frac{b_k(e^{i\theta}) + \sqrt{\delta_k(e^{i\theta})}}{2} \, d\theta.$$

Differentiating under the integral sign we have

$$\frac{d}{dk} \log \frac{b + \sqrt{b^2 - 4}}{2} = \frac{d}{db} \log \frac{b + \sqrt{b^2 - 4}}{2} \cdot \frac{db}{dk} = \frac{1}{\sqrt{b^2 - 4}},$$

hence

$$\frac{d\mu(k)}{dk} = \frac{1}{\pi} \, \mathrm{Re} \int_0^\pi \frac{d\theta}{\sqrt{\delta_k(e^{i\theta})}} = \frac{1}{\pi} \, \mathrm{Re} \int_{-1}^1 \frac{1}{\sqrt{(2u+k)^2 - 4}} \frac{du}{\sqrt{1 - u^2}},$$

where the change of variable $u = \cos\theta = (x + x^{-1})/2$ was implemented. The differentiation works for all $k > 0$ except $k = 4$, when the integral for $d\mu(k)/dk$ diverges. Finally, we change the variable

$$u = \frac{2(k - 2)t - k}{k - 4t} \quad \text{with } 0 < t < 1,$$

to reduce the integral to the desired form.   □

Lemma 3.10 *For real $k \geq 5$,*

$$\frac{d\tilde{\mu}(k)}{dk} = \frac{1}{\pi} \int_0^1 \frac{dt}{\sqrt{t(1 - t)(k^2 - 16t)}}.$$

Proof The change $X \mapsto X^2$ in the integral defining $\tilde{\mu}(k)$ does not affect the latter:

$$\tilde{\mu}(k) = \mathrm{m}(Y^2 + kY + (X + X^{-1})^2).$$

We write

$$Y^2 + kY + (X + X^{-1})^2 = (Y - Y_1(X))(Y - Y_2(X)),$$

where $Y_1(X) = -(k + \sqrt{\Delta(X)})/2$, $Y_2(X) = -(k - \sqrt{\Delta(X)})/2$, and $\Delta(X) = \Delta_k(X) = k^2 - 4(X + X^{-1})^2$ is real on $|X| = 1$ and positive for $k \geq 4$. Furthermore, for $k \geq 5$ we get $\Delta_k(X) \geq 5^2 - 4 \cdot 2^2 = 9$ and

$$|Y_1(X)| = \frac{k + \sqrt{\Delta(X)}}{2} \geq \frac{5 + 3}{2} = 4;$$

on using $|Y_1(X)Y_2(X)| = (X + X^{-1})^2 \leq 4$ we conclude that $|Y_2(X)| \leq 1$ for $k \geq 5$.

As in the previous proof we deduce that

$$\tilde{\mu}(k) = \frac{1}{2\pi i} \int_{|X|=1} \log|Y_1(X)| \frac{dX}{X} = \frac{1}{\pi} \int_0^\pi \log|Y_1(e^{i\theta})| \, d\theta$$

$$= \frac{1}{\pi} \int_0^\pi \log \frac{k + \sqrt{\Delta_k(e^{i\theta})}}{2} \, d\theta$$

implying

$$\frac{d\tilde{\mu}(k)}{dk} = \frac{1}{\pi} \int_0^\pi \frac{d\theta}{\sqrt{\Delta_k(e^{i\theta})}} = \frac{1}{\pi} \int_{-1}^1 \frac{du}{\sqrt{(1 - u^2)(k^2 - 16u^2)}}.$$

It remains to notice that

$$\int_{-1}^1 \frac{du}{\sqrt{(1 - u^2)(k^2 - 16u^2)}} = 2 \int_0^1 \frac{du}{\sqrt{(1 - u^2)(k^2 - 16u^2)}}$$

and perform the change of variable $u^2 = t$ in the result. □

Proof of Proposition 3.8 Since $P_k(x, y) = k + f(x, y)$, where $f(x, y)$ is absolutely bounded on the torus $|x| = |y| = 1$, we have

$$\mu(k) = \iint_{|x|=|y|=1} \log|k + f(x, y)| \frac{dx}{x} \frac{dy}{y}$$

$$= \iint_{|x|=|y|=1} (\log k + O(k^{-1})) \frac{dx}{x} \frac{dy}{y} = \log k + O(k^{-1}) \quad \text{as } k \to +\infty;$$

the same asymptotics is true for $\tilde{\mu}(k)$ in view of $\tilde{P}_k(X, Y)/Y = k + \tilde{f}(X, Y)$. It follows from Lemmas 3.9 and 3.10 that $\mu(k) - \tilde{\mu}(k)$ is constant for $k \geq 5$, which is seen to be zero from the asymptotics $\mu(k) - \tilde{\mu}(k) = O(k^{-1})$ as $k \to +\infty$. □

Chapter notes

A different proof of Proposition 3.1 and other related properties of the multivariate Mahler measure, including Theorem 3.7, are given in [176, Section 3.4]. Interestingly enough, in [130] Mahler himself does not touch explicitly on the convergence of the integral (3.1) defining the measure; however, Proposition 3.2 provides the needed upper and lower estimates for $M(P)$ in terms of the length of the polynomial P. To complete the proof of the well-definedness of the Mahler measure, one also has to show that for any non-zero polynomial $P(x_1, \ldots, x_k)$, the set $\{P = 0\} \cap \mathbb{T}^k$ has measure 0 in the torus, which can be done by induction on k. In fact, it is even possible to show that the set $\{|P| < \varepsilon\} \cap \mathbb{T}^k$ has measure $\ll \varepsilon^\delta$ for some constant $\delta > 0$ [83, Lemma 3.8], from which the convergence directly follows.

The convergence of Riemann sums to the Mahler measure for arbitrary multivariate polynomials was discussed by Lind, Schmidt and Verbitskiy [124], and quite recently Dimitrov and Habegger gave an estimate of the convergence rate in [72, Theorem 1.2]. At the same time, the approach via Riemann sums over the roots of unity, which we exploit in Section 3.4, can be used to define p-adic Mahler measures. This is done by Besser and Deninger in [24] (in the multivariate setting); for the single-variable case, their construction recovers the Shnirelman integral. Besser and Deninger also give another definition of p-adic Mahler measures using the p-adic syntomic regulator.

In relation to estimates given in Section 3.1, Mahler writes in [130] that 'it would ... have great interest to find the exact maxima' of the quotients

$$L(P_1 \cdots P_m)^{-1} \prod_{j=1}^{m} L(P_j) \quad \text{and} \quad H(P_1 \cdots P_m)^{-1} \prod_{j=1}^{m} H(P_j)$$

as functions of total degrees $d_j = \deg P_j$ for $j = 1, \ldots, m$, where the polynomials are from $\mathbb{C}[x_1, \ldots, x_k]$.

Theorems 3.5 and 3.7 suggest an interesting structure of the (countable!) set $L^{\#} = \{\mathrm{M}(P) : P \in \mathbb{Z}[x_1, \ldots, x_k]$ for some $k\} \subset [1, \infty)$. In his beautiful exposition [38], Boyd conjectures and gives evidence for $L^{\#}$ to be a closed set — the fact implying the answer to Lehmer's problem in the negative, in other words, the existence of $\varepsilon_0 > 0$ such that $\mathrm{M}(P(x)) \geq 1 + \varepsilon_0$ for all non-cyclotomic $P(x) \in \mathbb{Z}[x]$.

There is a related extension of Kronecker's theorem, proved independently by Lawton [120], Boyd [37] and Smyth [197]. It states that if the Mahler measure of a polynomial $P(x_1, \ldots, x_k) \in \mathbb{Z}[x_1, \ldots, x_k]$ is equal to 1, then it is a product of a finite number of factors of the form $x_1^{r_1} \cdots x_k^{r_k}$ and $\Phi(x_1^{s_1} \cdots x_k^{s_k})$, where $\Phi(x)$ is a cyclotomic polynomial (and all the exponents of monomials are integers).

The method in Section 3.5 for proving the coincidence of Mahler measures is sufficiently universal to deal with more sophisticated examples. In [22, 23] it is used for equating the Mahler measures of hyperelliptic polynomials to those of elliptic polynomials. We will also experience the power of the differentiation machinery for parametric families of Mahler measures in the next chapters.

Additional exercises

We present a method to determine whether a given polynomial $P(x_1, \ldots, x_k)$ vanishes on the torus $\mathbb{T}^k : |x_1| = \cdots = |x_k| = 1$. It is based on an idea of Rodriguez Villegas [62].

Exercise 3.8 Consider the Möbius transformation

$$M(z) = \frac{z - i}{z + i}.$$

(a) Verify that M maps the real line onto the unit circle deprived of 1.

(b) Let $P(x)$ be a complex polynomial of degree d. Show that the zeros of P in $\mathbb{T} \setminus \{1\}$ are in bijection with the real zeros of the polynomial $\tilde{P}(z) = (z + i)^d P(M(z))$.

(c) Let $P(x_1, \ldots, x_k)$ be a complex polynomial, and let Z_P be its zero locus in $\mathbb{C}^k$. Show that there exist explicit polynomials $A, B \in \mathbb{R}[z_1, \ldots, z_k]$ such that $Z_P \cap (\mathbb{T} \setminus \{1\})^k$ is in bijection with the set of common zeros of A and B in $\mathbb{R}^k$.

(d) In Section 3.3, we found the intersection of Z_P with the torus for the polynomials $P(x, y) = 1 + x + y$ and $P(x, y, z) = 1 + x + y + z$, respectively. Use the above method to recover these results.

(e) Find a necessary and sufficient condition on $k \in \mathbb{R}$ such that the Laurent polynomial

$$P(x, y) = x + \frac{1}{x} + y + \frac{1}{y} + xy + \frac{1}{xy} + \frac{x}{y} + \frac{y}{x} + k$$

vanishes on the torus.

Hint (e) The method gives a condition of the form $(t^2 + 1)(u^2 + 1)k = R(t, u)$ for some $t, u \in \mathbb{R}$ and some polynomial R. Set $\tilde{t} = 1/(t^2 + 1)$, $\tilde{u} = 1/(u^2 + 1)$. □

Exercise 3.9 ([159, 184]) For a *convex* lattice polygon N in $\mathbb{Z}^2$, the notation Area(N), bound(N) and int(N) are used for its area, the number of boundary and internal lattice points, respectively.

(a) Prove Pick's formula

$$\text{Area}(N) = \text{int}(N) + \frac{1}{2} \text{bound}(N) - 1.$$

(b) Show the inequality

$$\text{bound}(N) \leq 2\,\text{int}(N) + 7.$$

A goal of the next exercise is to characterise all one-parameter families of tempered reciprocal polynomials $P(x, y)$ whose Newton polygons N_P have exactly one internal point. The latter condition means that $P(x, y) = 0$ is (generically) an elliptic curve (see Section 3.2), while the temperedness implies that the parameter can only be assigned to the monomial corresponding to the internal point of N_P. One such example, $P(x, y) = x + 1/x + y + 1/y + k$, is treated in Section 3.5. Liu [125] computed a list of tempered families $P_k(x, y)$

with one internal point, as well as the *j*-invariants of the corresponding elliptic curves. This gives a complete list of possible *j*-invariants. The possible degrees of these rational functions of k are 4, 6, 8, 9 and 12, which means that not every elliptic curve over $\mathbb{Q}$ can appear in these families for rational values of k.

Exercise 3.10 (Branković [44]) (a) We say that the two convex polygons N and $\tilde{N}$ are affine equivalent if $\tilde{N} = gN + v$ for some $g \in GL_2(\mathbb{Z})$ and $v \in \mathbb{Z}^2$. Show that if N is a lattice triangle with exactly one internal lattice point, then it is equivalent to one of the five triangles in Figure 3.1.

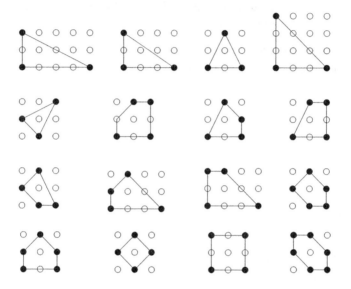

Figure 3.3 All non-equivalent convex lattice polygons with one internal lattice point

(b) Show that there are exactly 16 non-equivalent convex lattice polygons with one internal lattice point, reproduced in Figure 3.3.

(c) Prove that only 3 out of the 16 polygons from part (b) are centrally symmetric (these are the last instances in Figure 3.3).

(d) All one-parameter families of tempered reciprocal polynomials in $\mathbb{Z}[x, y]$ having one internal point, up to affine equivalence and changes $x \leftrightarrow y$ and/or $y \mapsto -y$, are exhausted by the following list:

$$x + \frac{1}{x} + \varepsilon_1\left(y + \frac{1}{y}\right) + k, \quad x + \frac{1}{x} + \varepsilon_1\left(y + \frac{1}{y}\right) + \varepsilon_2\left(\frac{x}{y} + \frac{y}{x}\right) + k,$$

$$xy + \frac{1}{xy} + \varepsilon_1\left(\frac{x}{y} + \frac{y}{x}\right) + \delta_1\left(x + \frac{1}{x}\right) + \delta_2\left(y + \frac{1}{y}\right) + k,$$

where $\varepsilon_1, \varepsilon_2 \in \{\pm 1\}$ and, for the last entry, $\delta_1 = \delta_2 = 0$ if $\varepsilon_1 \neq 1$ and $\delta_1, \delta_2 \in \{0, \pm 1, \pm 2\}$ if $\varepsilon_1 = 1$.

(e) Show that for each of the families in part (d) there is a birational transformation of the coordinates (x, y) such that the zero locus of its image is given by a tempered but non-reciprocal polynomial.

Hint   (a), (b) Use Exercise 3.9.

(e) In his thesis [44], Branković shows that in most cases one can take $x = (X^2 + 1)Y$, $y = 1/(XY)$ to break the reciprocal symmetry. This, however, increases the genus of the family. Is there always such a birational map that does not increase the genus?   □

Exercise 3.11 Give an example of a tempered polynomial $P(x, y)$ such that the polynomial $P(x - 1, y)$ is tempered as well.

4

The dilogarithm

As seen in the proof of Proposition 3.4, the dilogarithm function

$$\mathrm{Li}_2(z) = -\int_0^z \log(1-x)\,\frac{\mathrm{d}x}{x}, \tag{4.1}$$

better known through its series representation

$$\mathrm{Li}_2(z) = \sum_{n=1}^{\infty} \frac{z^n}{n^2} \tag{4.2}$$

for $|z| \le 1$, quite naturally shows up in the evaluation of a two-variate Mahler measure. This is not accidental — the goal of this section is to link more general two-variate Mahler measures with their dilogarithmic relative [27, 81, 219], the Bloch–Wigner dilogarithm

$$D(z) = \mathrm{Im}\,\mathrm{Li}_2(z) + \arg(1-z)\,\log|z|. \tag{4.3}$$

4.1 The q-binomial theorem and pentagonal identity

Throughout this section, we implement the q-Pochhammer symbol

$$(z)_n = (z;q)_n = \prod_{j=0}^{n-1}(1-zq^j) \quad \text{for } n = 1, 2, \ldots, \quad \text{and} \quad (z)_0 = 1,$$

where the parameter q assumes real values in the interval $0 < q < 1$. Note that n can also be taken to be infinite: when q is fixed in the interval, the product $(z)_\infty = (z;q)_\infty = \prod_{n=0}^{\infty}(1-zq^n)$ defines an entire function in $z \in \mathbb{C}$.

For $z \in \mathbb{C}$, $|z| < 1$, define the q-exponential function

$$e(z) = e_q(z) = \frac{1}{(z)_\infty} = \frac{1}{\prod_{n=0}^{\infty}(1-zq^n)}. \tag{4.4}$$

The similarity with the classical exponential function comes from the expansion

$$e_q(z) = \sum_{n=0}^{\infty} \frac{z^n}{(q)_n} = \sum_{n=0}^{\infty} \frac{(1-q)^n z^n}{[n]_q!}, \tag{4.5}$$

where the q-polynomials

$$[n]_q! = \frac{(q)_n}{(1-q)^n} = \prod_{k=1}^{n} \frac{1-q^k}{1-q}$$

may be viewed as q-factorials (that is, 'q-analogues' of the factorials) since $[n]_q! \to n!$ as $q \to 1$. In addition, this function satisfies the following 'exponential' functional identity due to Schützenberger.

Exercise 4.1 ([183]) Verify that

$$e(X + Y) = e(X)e(Y)$$

if $e(X) = e_q(X)$, $e(Y) = e_q(Y)$ and $e(X + Y) = e_q(X + Y)$ are viewed as elements in the algebra $\mathbb{C}_q[[X, Y]]$ of formal power series in two elements X, Y linked by the commutation relation $XY = qYX$.

On the other hand, from (4.4) we have the asymptotic behaviour

$$\log e(z) = \sum_{n=0}^{\infty} (-\log(1 - q^n z)) = \sum_{n=0}^{\infty} \sum_{m=1}^{\infty} \frac{q^{mn} z^m}{m} = \sum_{m=1}^{\infty} \frac{z^m}{m(1 - q^m)}$$

$$= \frac{1}{1-q} \sum_{m=1}^{\infty} \frac{z^m}{m(1 - q^m)/(1-q)} \sim \frac{-1}{\log q} \sum_{m=1}^{\infty} \frac{z^m}{m^2} \quad \text{as } q \to 1 \tag{4.6}$$

(already known to Ramanujan [17]), since $(1-q^m)/(1-q) \to m$ and $\log q \sim q-1$ as $q \to 1$. This allows (4.6) to be thought of as a q-analogue of the dilogarithm function (4.2). The function $\log e(z)$ is indeed known in the literature under the name 'quantum dilogarithm' (so as not to mix it up with the elliptic dilogarithm defined later in (4.30)). This analogy is much deeper than just the asymptotics above because of the following.

Exercise 4.2 ([110, 219]) Show that the q-binomial theorem [89]

$$\sum_{n=0}^{\infty} \frac{(x)_n}{(q)_n} y^n = \frac{(xy)_\infty}{(y)_\infty} \tag{4.7}$$

is equivalent to the so-called quantum pentagonal identity

$$e(X)e(Y) = e(Y)e(-YX)e(X), \tag{4.8}$$

where, as in Exercise 4.1, $e(X) = e_q(X)$, $e(Y) = e_q(Y)$ and $e(-YX) = e_q(-YX)$

are elements in the algebra $\mathbb{C}_q[[X, Y]]$ of formal power series in two elements X, Y linked by the commutation relation $XY = qYX$.

It seems that Richmond and Szekeres [162] were the first to realise that the limiting case $q \to 1$ of certain q-hypergeometric identities (actually, they considered the Andrews–Gordon generalisation of the Rogers–Ramanujan identities) produces non-trivial identities for the dilogarithm values; we return to this point in additional exercises to this chapter. The argument was later exploited by Loxton [128] and rediscovered in the context of (4.7), (4.8) by Faddeev and Kashaev [84, 110].

Below we review the derivation of the five-term (pentagonal) identity for the dilogarithm from the q-binomial theorem, in a somewhat different form compared to [110, 128, 162].

Theorem 4.1 *The dilogarithm function* (4.1) *satisfies the identity*

$$\mathrm{Li}_2(x) + \mathrm{Li}_2(y) = \mathrm{Li}_2\left(\frac{x}{1-y}\right) + \mathrm{Li}_2\left(\frac{y}{1-x}\right) - \mathrm{Li}_2\left(\frac{xy}{(1-x)(1-y)}\right)$$
$$- \log(1-x)\log(1-y). \tag{4.9}$$

Formula (4.9) is due to Abel [1] but an equivalent formula had been published by Spence [200] nearly twenty years earlier. Another equivalent form of (4.9) (recorded in (4.20) below) was given by Rogers [166].

Proposition 4.2 *Assume that* $0 < x, y < 1$. *Then the limiting case* $q \to 1$ *of the q-binomial theorem* (4.7) *is the equality*

$$\mathrm{Li}_2(x) + \mathrm{Li}_2\left(\frac{1-y}{1-xy}\right) - \mathrm{Li}_2(1) - \mathrm{Li}_2\left(x\frac{1-y}{1-xy}\right) + \log y \cdot \log \frac{1-y}{1-xy}$$
$$= \mathrm{Li}_2(xy) - \mathrm{Li}_2(y). \tag{4.10}$$

Although we prove relation (4.10) for x and y restricted to the interval $(0, 1)$, and this positivity is always crucial in application of the allied asymptotical formulae (cf. [128]), the identity remains valid for generic complex x, y by analytic continuation.

Proof Without loss of generality assume that q is sufficiently close to 1, namely, that

$$\max\{x, y, 1 - y(1 - x)\} < q < 1. \tag{4.11}$$

The easy part of the proposition is the asymptotics of the right-hand side in (4.7):

$$\log \frac{(xy)_\infty}{(y)_\infty} = \log \frac{e(y)}{e(xy)} \sim \frac{1}{\log q}(\mathrm{Li}_2(xy) - \mathrm{Li}_2(y)) \quad \text{as } q \to 1, \tag{4.12}$$

which is obtained on the basis of (4.6).

For the left-hand side of (4.7), write

$$\sum_{n=0}^{\infty} \frac{(x)_n}{(q)_n} y^n = \sum_{n=0}^{\infty} c_n, \quad \text{where } c_n = \frac{(x)_n}{(q)_n} y^n > 0. \tag{4.13}$$

Then the sequence

$$d_n = \frac{c_{n+1}}{c_n} = \frac{1 - xq^n}{1 - q^{n+1}} y > 0, \quad n = 0, 1, 2, \ldots, \tag{4.14}$$

satisfies

$$\frac{d_{n+1}}{d_n} = \frac{(1 - xq^{n+1})(1 - q^{n+1})}{(1 - xq^n)(1 - q^{n+2})} = 1 - \frac{q^n(1 - q)(q - x)}{(1 - xq^n)(1 - q^{n+2})}$$

$$< 1 - q^n(1 - q)(q - x), \quad n = 0, 1, 2, \ldots \tag{4.15}$$

(we use $0 < x < q < 1$), hence it is strictly decreasing. On the other hand, $1 - y(1 - x) < q$ implies

$$d_0 = \frac{c_1}{c_0} = \frac{1 - x}{1 - q} y > 1,$$

while

$$\lim_{n \to \infty} d_n = \lim_{n \to \infty} \frac{1 - xq^n}{1 - q^{n+1}} y = y < 1;$$

thus, there exists a unique index $N \geq 1$ such that

$$d_{N-1} = \frac{c_N}{c_{N-1}} \geq 1 \quad \text{and} \quad d_N = \frac{c_{N+1}}{c_N} < 1. \tag{4.16}$$

Solving the inequality $c_{n+1}/c_n < 1$, or equivalently $(1 - xq^n)y < 1 - q^{n+1}$, we obtain $n > T$, where

$$T = \frac{1}{\log q} \cdot \log \frac{1 - y}{q - xy}, \tag{4.17}$$

hence $N = \lfloor T \rfloor$. From (4.14)–(4.16) we conclude that c_N is the main term contributing the sum in (4.13), namely,

$$c_N < \sum_{n=0}^{\infty} c_n < c_N \times \text{const}.$$

This implies

$$\log \sum_{n=0}^{\infty} c_n \sim \log c_N = \log\left(\frac{e(q)e(xq^N)}{e(x)e(q^{N+1})} y^N\right)$$

$$\sim \log\left(\frac{e(q)e(xq^T)}{e(x)e(q^{T+1})} y^T\right) \quad \text{as } q \to 1. \tag{4.18}$$

Note now that from (4.17) we have

$$q^T = \frac{1-y}{q-xy},$$

whence the asymptotics in (4.18) may be continued as follows:

$$\log \sum_{n=0}^{\infty} c_n \sim \log e(q) + \log e\left(x\frac{1-y}{q-xy}\right) - \log e(x) - \log e\left(q\frac{1-y}{q-xy}\right)$$
$$+ \frac{\log y}{\log q} \cdot \log \frac{1-y}{q-xy}$$
$$\sim \frac{1}{\log q}\left(\mathrm{Li}_2(x) + \mathrm{Li}_2\left(\frac{1-y}{1-xy}\right) - \mathrm{Li}_2(1) - \mathrm{Li}_2\left(x\frac{1-y}{1-xy}\right)\right)$$
$$+ \log y \cdot \log \frac{1-y}{1-xy}\right) \qquad \text{as } q \to 1, \tag{4.19}$$

where (4.6) is used.

Comparing the asymptotics (4.12) and (4.19) of both sides of (4.7) we arrive at the identity (4.10).   □

Exercise 4.3 (a) Show that (4.10) implies

$$\mathrm{Li}_2(xy) + \mathrm{Li}_2\left(\frac{x(1-y)}{1-xy}\right) + \mathrm{Li}_2\left(\frac{y(1-x)}{1-xy}\right) + \log\frac{1-y}{1-xy} \cdot \log\frac{1-x}{1-xy}$$
$$= \mathrm{Li}_2(x) + \mathrm{Li}_2(y). \tag{4.20}$$

(b) Prove the equivalence of the functional equations (4.20) and (4.9).

In other words, Proposition 4.2 implies Theorem 4.1.

Hint (a) Take $x = 0$ in (4.10) to get

$$\mathrm{Li}_2(y) + \mathrm{Li}_2(1-y) - \mathrm{Li}_2(1) + \log y \cdot \log(1-y) = 0.$$

This identity, in particular, implies

$$\mathrm{Li}_2\left(\frac{1-y}{1-xy}\right) - \mathrm{Li}_2(1) = -\mathrm{Li}_2\left(1 - \frac{1-y}{1-xy}\right) - \log\frac{1-y}{1-xy} \cdot \log\left(1 - \frac{1-y}{1-xy}\right).$$

Substitute this result into (4.10) to get (4.20).

(b) Change variables $\tilde{x} = x(1-y)/(1-xy)$, $\tilde{y} = y(1-x)/(1-xy)$ in (4.20).   □

4.2 The Rogers and Bloch–Wigner dilogarithms

The dilogarithm (4.1) satisfies a second-order (inhomogeneous) linear differential equation with (regular) singularities at 0, 1 and ∞. The latter two points are log singularities of the function, so that defining it globally as a complex-analytic function requires excluding from $\mathbb{C}$ a cut joining 1 and ∞. There are, however, real-analytic functions with lesser restrictions on the analyticity domain and a simpler form of the five-term relation (4.9).

The Rogers dilogarithm [166, 219] is a real function defined in the interval $0 < x < 1$ by the formula

$$L(x) = \mathrm{Li}_2(x) + \frac{1}{2} \log x \log(1 - x) = \sum_{n=1}^{\infty} \frac{x^n}{n^2} - \frac{1}{2} \log x \sum_{n=1}^{\infty} \frac{x^n}{n} \qquad (4.21)$$

and then extended to the rest of the real line by setting $L(0) = 0$, $L(1) = \pi^2/6$, and

$$L(x) = \begin{cases} 2L(1) - L\left(\dfrac{1}{x}\right) & \text{if } x > 1, \\ -L\left(\dfrac{-x}{1 - x}\right) & \text{if } x < 0. \end{cases} \qquad (4.22)$$

The resulting function is then a monotone increasing continuous real-valued function on $\mathbb{R}$ with limiting values

$$\lim_{x \to -\infty} L(x) = -L(1) = -\frac{\pi^2}{6} \quad \text{and} \quad \lim_{x \to +\infty} L(x) = 2L(1) = \frac{\pi^2}{3}. \qquad (4.23)$$

Furthermore, it is real analytic except at $x = 0$ and $x = 1$, where its derivative becomes infinite. The five-term relation (4.20) for (4.21), (4.22) reads [166]

$$L(xy) + L\left(\frac{x(1 - y)}{1 - xy}\right) + L\left(\frac{y(1 - x)}{1 - xy}\right) = L(x) + L(y). \qquad (4.24)$$

Some particular cases of the latter are

$$L(x) + L(1 - x) = L(1) \qquad (4.25)$$

and Abel's duplication formula

$$L(x^2) = 2L(x) - 2L\left(\frac{x}{1 + x}\right). \qquad (4.26)$$

Richmond and Szekeres [162] realised that the Rogers dilogarithm is one of the most appropriate functions to express the limiting case $q \to 1$ of q-identities; one can also see this by comparing the 'cleaned' identity (4.24) with (4.20).

The function $\mathrm{Li}_2(z)$ in (4.1) is complex-analytic on $\mathbb{C} \setminus [1, \infty)$; crossing the cut $[1, \infty)$ it jumps by $2\pi i \log |z|$. This means that the 'correction' $\mathrm{Li}_2(z) + i \arg(1 - z) \log |z|$, where the branch of arg is chosen with values between $-\pi$ and π, is continuous on $\mathbb{C}$. Surprisingly enough, the imaginary part (4.3) of this correction, known as the Bloch–Wigner dilogarithm, is not just continuous (as a real-valued function) on $\mathbb{C}$ but satisfies stronger analyticity properties.

Exercise 4.4 ([27, 219]) (a) Show that $D(z)$ is real analytic on $\mathbb{C} \setminus \{0, 1\}$.
(b) Verify that $D(z)$ is not differentiable at 0 and 1.
(c) Show that the differential of $D(z)$ at $z \in \mathbb{C} \setminus \{0, 1\}$ is given by

$$dD(z) = \log |z| \, d \arg(1 - z) - \log |1 - z| \, d \arg(z),$$

where $d \arg(h) = \operatorname{Im}(d \log h) = \operatorname{Im}(dh/h)$.

Similar to the case of the Rogers dilogarithm $L(x)$, the five-term relation(s) for the Bloch–Wigner dilogarithm (4.3) lose the logarithmic terms; in particular, Theorem 4.1 reduces to

$$D(x) + D(y) - D\left(\frac{x}{1 - y}\right) - D\left(\frac{y}{1 - x}\right) + D\left(\frac{xy}{(1 - x)(1 - y)}\right) = 0. \qquad (4.27)$$

Exercise 4.5 (a) Give details of the proof of (4.27).
(b) Show that $D(1/z) = -D(z)$ and $D(1-z) = -D(z)$, so that the Bloch–Wigner dilogarithm admits the 6-fold symmetry

$$D(z) = D\left(1 - \frac{1}{z}\right) = D\left(\frac{1}{1 - z}\right)$$

$$= -D\left(\frac{1}{z}\right) = -D(1 - z) = -D\left(\frac{-z}{1 - z}\right).$$

The exercise implies that

$$D(0) = D(1) = D(\infty) = 0, \quad \text{where } D(\infty) := \lim_{z \to \infty} D(z).$$

Another important property of the function (4.3), due to Kummer, is that its values can always be expressed in terms of the Clausen function

$$D(e^{2\pi i t}) = \operatorname{Im} \operatorname{Li}_2(e^{2\pi i t}) = \sum_{n=1}^{\infty} \frac{\sin 2\pi n t}{n^2}$$

of a *real* variable t, which in turn links to the Lobachevsky function

$$Л(\theta) = -\int_0^{\theta} \log |2 \sin u| \, du = \frac{1}{2} \sum_{n=1}^{\infty} \frac{\sin 2n\theta}{n^2}$$

for computing hyperbolic volumes (see Milnor's Chapter 7 in Thurston's lecture notes [208]).

Exercise 4.6 (Kummer's relation) Show that

$$D(z) = \frac{1}{2}\left(D\left(\frac{z}{\bar{z}}\right) + D\left(\frac{1 - 1/z}{1 - 1/\bar{z}}\right) + D\left(\frac{1 - \bar{z}}{1 - z}\right)\right).$$

A consequence of the formula is that $D(z)$ can be thought of as a continuous function on $\mathbb{P}^1(\mathbb{C})$, which is real analytic on $\mathbb{P}^1(\mathbb{C}) \setminus \{0, 1, \infty\}$.

4.3 Maillot's formula

We will see later in Chapter 7 that the Bloch–Wigner dilogarithm (4.3) shows up naturally in the geometry of Mahler-measure evaluations. In this section we limit ourselves to a simple illustration of the principles to come: we give a foretaste of the story through a particular entry known as Maillot's formula [132] (also as the Cassaigne–Maillot formula) which generalises Smyth's evaluation from Proposition 3.4 and possesses an elegant plane-geometry formulation. One can also view it as an extension of Jensen's formula $m(ax + b) = \log \max\{|a|, |b|\}$ to two variables.

Proposition 4.3 *Given $a, b, c \in \mathbb{C}$, assume that $|a|, |b|$ and $|c|$ form the sides of a non-degenerate plane triangle, whose corresponding opposite angles are α, β and γ. Then*

$$m(ax + by + c) = \frac{1}{\pi}\left(\alpha \log|a| + \beta \log|b| + \gamma \log|c| + D\left(\frac{|b|}{|a|} e^{i\gamma}\right)\right), \qquad (4.28)$$

where $D(\,\cdot\,)$ is the Bloch–Wigner dilogarithm.

If no triangle with sides $|a|, |b|$ and $|c|$ exists, then

$$m(ax + by + c) = \log \max\{|a|, |b|, |c|\}.$$

Let $A, B, C \in \mathbb{C}$ be vertices of a positively oriented plane triangle with sides $|a|, |b|$ and $|c|$, so that α, β and γ are the corresponding adjacent angles, $\alpha + \beta + \gamma = \pi$. One way of seeing that the right-hand side in (4.28) is invariant under cyclic permutations of $|a|, |b|$ and $|c|$ (equivalently, of A, B and C) is to associate the value

$$D(ABC) = D\left(\frac{A - B}{C - B}\right).$$

Then $D(ABC) = D(BCA) = D(CAB)$ by Exercise 4.5(b), while Exercise 4.7 leads to an equivalent form

$$m(ax + by + c) = \frac{1}{\pi}(\alpha \log|a| + \beta \log|b| + \gamma \log|c| + D(ABC))$$

of (4.28).

Exercise 4.7 In the notation above, show that

$$D\left(\frac{|b|}{|a|} e^{i\gamma}\right) = D(ABC).$$

Proof of Proposition 4.3 First notice that

$$m(ax + by + c) = m(ax + by + cz)$$

$$= \frac{1}{(2\pi)^3} \iiint_{[-\pi,\pi]^3} \log\left|ae^{i\theta_1} + be^{i\theta_2} + ce^{i\theta_3}\right| d\theta_1 \, d\theta_2 \, d\theta_3$$

$$= \frac{1}{(2\pi)^3} \iiint_{[-\pi,\pi]^3} \log \left| |a|e^{i\theta_1} + |b|e^{i\theta_2} + |c|e^{i\theta_3} \right| d\theta_1 \, d\theta_2 \, d\theta_3$$

$$= \mathrm{m}(|a|x + |b|y + |c|z) = \mathrm{m}(|a|x + |b|y + |c|),$$

so that the result depends only on the absolute values of complex quantities a, b and c. This reduces verification to the case when these parameters are real and positive, which we will assume from now on. Furthermore, without loss of generality, the three numbers can be ordered — a consequence of Exercise 4.7.

Given $a \geq b \geq c > 0$, Jensen's formula gives the following reduction of the two-variate Mahler measure:

$$\mathrm{m}(ax + by + c) = \frac{1}{2\pi} \int_{-\pi}^{\pi} \log \max\{|a + be^{i\theta}|, c\} \, d\theta.$$

If no triangle with sides a, b and c exists then $a \geq b + c$, and vice versa; in that case $|a + be^{i\theta}| \geq a - b \geq c$ and

$$\frac{1}{2\pi} \int_{-\pi}^{\pi} \log \max\{|a + be^{i\theta}|, c\} \, d\theta = \frac{1}{2\pi} \int_{-\pi}^{\pi} \log |a + be^{i\theta}| \, d\theta = \log \max\{a, b\} = a,$$

again thanks to Jensen's formula. If a triangle exists, then $|a + be^{i\varphi}| = c$ for $\varphi = \pi - \gamma = \alpha + \beta$, and computation brings us to

$$\mathrm{m}(ax + by + c) = \frac{1}{\pi} \int_{\varphi}^{\pi} \log c \, d\theta + \frac{1}{2\pi} \int_{-\varphi}^{\varphi} \log |a + be^{i\theta}| \, d\theta$$

$$= \frac{\gamma}{\pi} \log c + \frac{\alpha + \beta}{\pi} \log a + \frac{1}{2\pi} \int_{-\varphi}^{\varphi} \log \left| 1 + \frac{b}{a} e^{i\theta} \right| \, d\theta.$$

For the remaining integral, we have

$$\frac{1}{2\pi} \int_{-\varphi}^{\varphi} \log |1 + re^{i\theta}| \, d\theta = \frac{1}{2\pi} \, \mathrm{Re} \int_{-\varphi}^{\varphi} \log(1 + re^{i\theta}) \, d\theta$$

$$= \frac{1}{2\pi} \, \mathrm{Im} \int_{-re^{-i\varphi}}^{re^{-i\varphi}} \log(1 - x) \, \frac{dx}{x}$$

$$= \frac{1}{2\pi} \, \mathrm{Im}(\mathrm{Li}_2(-re^{-i\varphi}) - \mathrm{Li}_2(-re^{-i\varphi}))$$

$$= \frac{1}{2\pi} \, \mathrm{Im} \, \mathrm{Li}_2(re^{i\gamma}) - \frac{1}{2\pi} \, \mathrm{Im} \, \mathrm{Li}_2(re^{-i\gamma})$$

$$= \frac{1}{\pi} \, \mathrm{Im} \, \mathrm{Li}_2(re^{i\gamma})$$

$$= \frac{1}{\pi} D(re^{i\gamma}) - \frac{1}{\pi} \log r \, \mathrm{Im} \log(1 - re^{i\gamma}),$$

where $r = b/a$. Finally, notice that $\mathrm{Im} \log(1 - re^{i\gamma}) = \mathrm{Im} \log(a - be^{i\gamma}) = -\beta$. ☐

Chapter notes

It is classical (see, for example, [217, Proposition 10.1]) that the cross ratio

$$\lambda(z_1, z_2, z_3, z_4) = \frac{(z_1 - z_4)(z_2 - z_3)}{(z_1 - z_3)(z_2 - z_4)}$$

gives a realisation of the configuration space of four points on the projective line $\mathbb{P}^1(\mathbb{C})$. Thanks to the symmetry from Exercise 4.5(b), the composition of the cross ratio with the Bloch–Wigner dilogarithm

$$\tilde{D}(z_1, z_2, z_3, z_4) = D\left(\frac{(z_1 - z_4)(z_2 - z_3)}{(z_1 - z_3)(z_2 - z_4)}\right) \tag{4.29}$$

respects the permutations σ of the four variables:

$$\tilde{D}(z_{\sigma(1)}, z_{\sigma(2)}, z_{\sigma(3)}, z_{\sigma(1)}) = \text{sign}(\sigma)\, \tilde{D}(z_1, z_2, z_3, z_4).$$

One can check then that the five-term relation for $D(z)$ can be stated equivalently in the form

$$\tilde{D}(z_1, z_2, z_3, z_4) + \tilde{D}(z_2, z_3, z_4, z_5) + \tilde{D}(z_3, z_4, z_5, z_1) + \tilde{D}(z_4, z_5, z_1, z_2)$$
$$+ \tilde{D}(z_5, z_1, z_2, z_3) = 0 \quad \text{for all } z_1, z_2, z_3, z_4, z_5 \in \mathbb{P}^1(\mathbb{C}).$$

The quantity (4.29) has a simple geometric interpretation of the hyperbolic volume of the tetrahedron with vertices $z_1, z_2, z_3, z_4 \in \mathbb{P}^1(\mathbb{C})$ [219]. With this in mind, the five-term relation for $\tilde{D}(\,\cdot\,)$ represents the fact that the five participating tetrahedra add up algebraically to the zero 3-cycle.

In [210, Theorem 6], Vandervelde gives a generalisation of the Cassaigne–Maillot formula for computing the Mahler measure $m(axy + bx + cy + d)$. Both this result and Proposition 4.3 are particular cases of $m(P(x, y))$, for which the curve $C = \{(x, y) \in \mathbb{C}^2 : P(x, y) = 0\}$ has genus 0. A generalisation of such a result to genus 1 (elliptic) curves brings the Bloch elliptic dilogarithm [27]

$$D_q(x) = \sum_{n=-\infty}^{\infty} D(xq^n), \quad \text{where } q = e^{2\pi i \tau}, \tag{4.30}$$

into play. Here the elliptic curve $\mathbb{C}/(\mathbb{Z} + \mathbb{Z}\tau)$ is identified with the quotient $\mathbb{C}^\times / q^{\mathbb{Z}}$ via the exponential map, so that $|q| < 1$.

Additional exercises

The aim of the next exercise is to establish generalised Rogers–Ramanujan identities, whose asymptotics as $q \to 1^-$ is a natural source for certain relations between the values of dilogarithmic function.

Exercise 4.8 (Bressoud [45]) We stay with the q-Pochhammer notation $(z)_n = (z; q)_n$ of Section 4.1 and define the q-binomial coefficients $\begin{bmatrix} m \\ n \end{bmatrix} = \begin{bmatrix} m \\ n \end{bmatrix}_q$ (also known as the Gaussian polynomials) to be

$$\begin{bmatrix} m \\ n \end{bmatrix} = \frac{(q)_m}{(q)_n (q)_{m-n}} \quad \text{if } 0 \le n \le m$$

and 0 if $n < 0$ or $n > m$. Clearly,

$$\lim_{q \to 1} \begin{bmatrix} m \\ n \end{bmatrix}_q = \binom{m}{n}.$$

(a) Verify the following q-analogue of Pascal's triangle relations:

$$\begin{bmatrix} m \\ n \end{bmatrix} = \begin{bmatrix} m-1 \\ n \end{bmatrix} + q^{m-n} \begin{bmatrix} m-1 \\ n-1 \end{bmatrix}.$$

This, in particular, implies that $\begin{bmatrix} m \\ n \end{bmatrix}$ are indeed polynomials in $\mathbb{Z}[q]$.

(b) Use the q-binomial theorem (4.7) to prove

$$\sum_{n=0}^{m} x^n q^{n(n+1)/2} \begin{bmatrix} m \\ n \end{bmatrix} = (1 + xq)(1 + xq^2) \cdots (1 + xq^m).$$

This can alternatively be written in the convenient form

$$\sum_{n \in \mathbb{Z}} x^n q^{n(n+1)/2} \begin{bmatrix} m \\ n \end{bmatrix} = (-xq)_m,$$

because the participating q-binomial coefficients have finite support on $\mathbb{Z}$.

(c) Show Cauchy's identity

$$\sum_{n=-N}^{N} x^n q^{n(n+1)/2} \begin{bmatrix} 2N \\ N-n \end{bmatrix} = \prod_{j=1}^{N} (1 + xq^j)(1 + x^{-1}q^{j-1}).$$

(d) By taking the limit as $N \to \infty$ in the latter equality, arrive at the Jacobi triple product identity

$$\sum_{n \in \mathbb{Z}} x^n q^{n(n+1)/2} = \prod_{j=1}^{\infty} (1 - q^j)(1 + xq^j)(1 + x^{-1}q^{j-1}).$$

(e) For positive integers k and N prove that

$$(q)_{2N}^{-1} \sum_{n=-N}^{N} x^n q^{n((2k+1)n+1)/2} \begin{bmatrix} 2N \\ N-n \end{bmatrix}$$

$$= \sum_{N \geq n_1 \geq n_2 \geq \cdots \geq n_k \geq 0} \frac{q^{n_1^2 + n_2^2 + \cdots + n_k^2}}{(q)_{N-n_1}(q)_{n_1-n_2} \cdots (q)_{n_{k-1}-n_k}(q)_{2n_k}}$$

$$\times \prod_{j=1}^{n_k} (1 + xq^j)(1 + x^{-1}q^{j-1}).$$

(f) By specialising to $x = -1$ deduce from the latter identity that

$$(q)_{2N}^{-1} \sum_{n \in \mathbb{Z}} (-1)^n q^{kn+n(n+1)/2} \begin{bmatrix} 2N \\ N-n \end{bmatrix}$$

$$= \sum_{N \geq n_1 \geq n_2 \geq \cdots \geq n_{k-1} \geq 0} \frac{q^{n_1^2 + n_2^2 + \cdots + n_{k-1}^2}}{(q)_{N-n_1}(q)_{n_1-n_2} \cdots (q)_{n_{k-2}-n_{k-1}}(q)_{n_{k-1}}}.$$

(g) Prove that

$$\sum_{n_1 \geq n_2 \geq \cdots \geq n_{k-1} \geq 0} \frac{q^{n_1^2 + n_2^2 + \cdots + n_{k-1}^2}}{(q)_{n_1-n_2} \cdots (q)_{n_{k-2}-n_{k-1}}(q)_{n_{k-1}}}$$

$$= \frac{(q^k; q^{2k+1})_\infty (q^{k+1}; q^{2k+1})_\infty (q^{2k+1}; q^{2k+1})_\infty}{(q; q)_\infty}. \tag{4.31}$$

These are particular instances of the Andrews–Gordon identities; others can be achieved by specialising to $x = -q^\ell$ for $\ell = 1, 2, \ldots, k-1$ in the identity in part (e). The Rogers–Ramanujan identities are the $k = 2$ instances:

$$\sum_{n=0}^{\infty} \frac{q^{n^2}}{(q; q)_n} = \frac{(q^2; q^5)_\infty (q^3; q^5)_\infty (q^5; q^5)_\infty}{(q; q)_\infty} = \prod_{j=0}^{\infty} \frac{1}{(1 - q^{5j+1})(1 - q^{5j+4})},$$

$$\sum_{n=0}^{\infty} \frac{q^{n^2+n}}{(q; q)_n} = \frac{(q; q^5)_\infty (q^4; q^5)_\infty (q^5; q^5)_\infty}{(q; q)_\infty} = \prod_{j=0}^{\infty} \frac{1}{(1 - q^{5j+2})(1 - q^{5j+3})}.$$

Hint   (c) Substitute xq^{-N} for x and $2N$ for m in the identity in part (b).

(e) This is performed by induction on k with Cauchy's identity from part (c) corresponding to $k = 0$ and serving as the base of induction. Details can be found in [45].

(g) Take the limit as $N \to \infty$ in part (f) and use the Jacobi triple product from part (d) with q replaced by q^{2k+1} and specialised to $x = -q^{-k}$. $\square$

Exercise 4.9 ([162, 216]) (a) Use the method from Section 4.1 to deduce from (4.31) the identity

$$\sum_{j=1}^{k-1} L\left(\frac{\sin^2 \frac{\pi}{2k+1}}{\sin^2 \frac{(j+1)\pi}{2k+1}}\right) = \frac{\pi^2(k-1)}{3(2k+1)}$$

for $k = 2, 3, \ldots$. Here $L(x)$ denotes the Rogers dilogarithm (4.21).

(b) Show that $L((\sqrt{5}-1)/2) = \pi^2/10$.

(c) Compute $D((\sqrt{5}-1)/2)$, where D denotes the Bloch–Wigner dilogarithm.

Hint (b) Use part (a) for $k = 2$ and relation (4.25) (or (4.26)). □

Exercise 4.10 ([209] and [32, Section 8]) Show that

(a) $m(y^2(x+1)^2 + y(x^2 + 6x + 1) + (x+1)^2) = \dfrac{8}{3} L'(\chi_{-4}, -1),$

(b) $m(y^2(x+1)^2 + y(x^2 - 10x + 1) + (x+1)^2) = \dfrac{20}{3} L'(\chi_{-3}, -1).$

5

Differential equations
for families of Mahler measures

In Chapter 3 we witnessed some parametric polynomial families, for which the Mahler measure differentiated with respect to the parameter is identified with a period on a related family of elliptic curves. The period viewed as a function of the parameter happens to satisfy a homogeneous linear differential equation of second order (and of a very special arithmetic type, known as a Picard–Fuchs equation). This means that the Mahler measure itself satisfies a linear differential equation. In this chapter we explore the latter property in greater detail, with the purpose of giving closed-form expressions for parametric Mahler measures in terms of hypergeometric functions and binomial sums.

5.1 A poly family

In this section, we are first interested in

$$\mu_q(k) = m((x_1 + x_1^{-1}) \cdots (x_q + x_q^{-1}) + k) = \mu_q(-k),$$

where $k > 0$ is real. When $q = 1$ this appears to be our baby example $\lambda^+(k)$. When $q = 2$,

$$(x_1 + x_1^{-1})(x_2 + x_2^{-1}) = x + x^{-1} + y + y^{-1}, \quad \text{where } x = x_1 x_2, \ y = x_1/x_2,$$

so that $\mu_2(k) = \mu(k)$ for the Mahler measure considered in Section 3.5. From this perspective, the family generalises those early examples. In addition, our strategy from Section 3.5 reasonably easily extends to the following statement.

Lemma 5.1 *For $q = 2, 3, \ldots$ and $k > 0$, with $k \neq 4$ when $q = 2$, we have*

$$\frac{d\mu_q(k)}{dk} = \frac{1}{\pi^{q-1}} \operatorname{Re} \int \cdots \int_{[0,1]^{q-1}} \frac{\prod_{j=1}^{q-1} t_j^{-1/2} (1 - t_j)^{-1/2} dt_j}{(k^2 - 4^q t_1 \cdots t_{q-1})^{1/2}}.$$

Proof The Laurent polynomial in question viewed as a polynomial in x_q, in which case it can be given as $bx_q + k + bx_q^{-1}$ where $b = b(x_1, \ldots, x_{q-1}) = \prod_{j=1}^{q-1}(x_j + x_j^{-1})$ assumes real values on the torus $|x_1| = \cdots = |x_{q-1}| = 1$, has two zeros whose product is 1. Therefore, the absolute value of one of the zeros is always at least 1, namely,

$$\frac{|k + \sqrt{k^2 - 4b^2}|}{2|b|} \geq 1,$$

while it is at most 1 for the other zero. This way we find that

$$\mu_q(k) = \frac{1}{(2\pi i)^{q-1}} \int \cdots \int_{|x_1| = \cdots = |x_{q-1}| = 1} \log\left|\frac{k + \sqrt{k^2 - 4b^2}}{2b}\right| \frac{dx_1}{x_1} \cdots \frac{dx_{q-1}}{x_{q-1}}$$

$$= \frac{1}{\pi^{q-1}} \operatorname{Re} \int \cdots \int_{[0,\pi]^{q-1}} \log \frac{k + \sqrt{k^2 - 4b^2}}{2b} d\theta_1 \cdots d\theta_{q-1},$$

where $b = b(e^{i\theta_1}, \ldots, e^{i\theta_{q-1}})$. Differentiation with respect to k under the integral sign followed by the change $u_j = \cos\theta_j$ for $j = 1, \ldots, q - 1$ results in

$$\frac{d\mu_q(k)}{dk} = \frac{1}{\pi^{q-1}} \operatorname{Re} \int \cdots \int_{[0,\pi]^{q-1}} \frac{1}{\sqrt{k^2 - 4b(e^{i\theta_1}, \ldots, e^{i\theta_{q-1}})^2}} d\theta_1 \cdots d\theta_{q-1}$$

$$= \frac{1}{\pi^{q-1}} \operatorname{Re} \int \cdots \int_{[-1,1]^{q-1}} \frac{1}{\sqrt{k^2 - 4^q(u_1 \cdots u_{q-1})^2}} \prod_{j=1}^{q-1} \frac{du_j}{\sqrt{1 - u_j^2}}$$

$$= \frac{2^{q-1}}{\pi^{q-1}} \operatorname{Re} \int \cdots \int_{[0,1]^{q-1}} \frac{du_1 \cdots du_{q-1}}{\sqrt{(1 - u_1^2) \cdots (1 - u_{q-1}^2)(k^2 - 4^q(u_1 \cdots u_{q-1})^2)}}.$$

Finally, change $t_j = u_j^2$ for $j = 1, \ldots, q - 1$. $\qquad\square$

The result of Lemma 5.1 can be stated as

$$\frac{d\mu_q(k)}{dk} = \frac{1}{\pi^{q-1}k} \operatorname{Re} F_q\left(\frac{4^q}{k^2}\right),$$

where

$$F_q(z) = \int \cdots \int_{[0,1]^{q-1}} \frac{\prod_{j=1}^{q-1} t_j^{-1/2}(1 - t_j)^{-1/2} dt_j}{(1 - zt_1 \cdots t_{q-1})^{1/2}}. \tag{5.1}$$

The integral in (5.1) makes sense for any $z \in \mathbb{C} \setminus [1, +\infty)$ and depends on the choice of branch for the surd when z is real, $z > 1$ (the choice affects its imaginary part only). The integral converges absolutely in the domain $\mathbb{C} \setminus [1, +\infty)$ and defines there an analytic function (see, for example, a very general statement in [147, Lemma 2]). When $|z| < 1$, we can use the binomial expansion

$$(1 - zt)^{-1/2} = \sum_{n=0}^{\infty} \frac{(\frac{1}{2})_n}{n!} z^n t^n,$$

where $(a)_n = a(a+1)\cdots(a+n-1) = \Gamma(a+n)/\Gamma(a)$ for $n = 0, 1, \ldots$ stands for the Pochhammer symbol (also known as the shifted factorial), to obtain

$$
\begin{aligned}
F_q(z) &= \sum_{n=0}^{\infty} \frac{(\frac{1}{2})_n}{n!} z^n \prod_{j=1}^{q-1} \int_0^1 t_j^{n-1/2}(1-t_j)^{-1/2} dt_j \\
&= \sum_{n=0}^{\infty} \frac{(\frac{1}{2})_n}{n!} z^n \left(\frac{\Gamma(n+\frac{1}{2})\Gamma(\frac{1}{2})}{\Gamma(n+1)} \right)^{q-1} = \pi^{q-1} \sum_{n=0}^{\infty} \frac{(\frac{1}{2})_n^q}{n!^q} z^n.
\end{aligned}
$$

The series we arrive at is a particular instance of the generalised hypergeometric series [9, 194]

$$
{}_qF_{q-1}\left(\begin{matrix} a_1, a_2, \ldots, a_q \\ b_2, \ldots, b_q \end{matrix} \,\middle|\, z \right) = \sum_{n=0}^{\infty} \frac{(a_1)_n(a_2)_n \cdots (a_q)_n}{n!\,(b_2)_n \cdots (b_q)_n} z^n.
$$

It satisfies a linear homogeneous differential equation with singularities at $z = 0$, 1 and ∞, and this differential equation provides an efficient way to continue the series analytically from the unit circle to $\mathbb{C}$ with the exclusion of the intervals $(-\infty, 0]$, $[0, 1]$ or $[1, \infty)$. The result of any such analytic continuation is called the generalised hypergeometric function, attached to the same set of parameters $a_1, \ldots, a_q$ and $b_2, \ldots, b_q$. (The details of the process of analytic continuation in the case $q = 2$ can be found in [217, Chapter III].)

Exercise 5.1 Verify that the coefficient $C(n)$ of the generalised hypergeometric series satisfies the recurrence equation

$$
C(n+1) = \frac{(n+a_1)(n+a_2)\cdots(n+a_q)}{(n+1)(n+b_2)\cdots(n+b_q)} C(n) \quad \text{for } n = 0, 1, 2, \ldots.
$$

Exercise 5.2 Define the differential operator

$$
\delta = z \frac{d}{dz}.
$$

Show that the following results are valid.

(a) For any polynomial $f(x) \in \mathbb{C}[x]$ and $n \in \mathbb{Z}$,

$$
f(\delta)z^n = f(n)z^n.
$$

(b) The operator δ does not commute with the operator (of multiplication by) z^s where $s \in \mathbb{R}$; but we have $\delta z^s = z^s(\delta + s)$.

(c) The generalised hypergeometric series $F(z) = {}_qF_{q-1}(z)$ satisfies the differential equation

$$
\left((\delta + a_1)(\delta + a_2)\cdots(\delta + a_q) - (\delta + 1)(\delta + b_1)\cdots(\delta + b_q)z^{-1} \right)F(z) = 0,
$$

or equivalently,

$$((\delta + a_1)(\delta + a_2) \cdots (\delta + a_q) - z^{-1}\delta(\delta + b_1 - 1) \cdots (\delta + b_q - 1))F(z) = 0.$$

In this last form $z^{-1}\delta = d/dz$.

Hint   (c) Use the recursion from Exercise 5.1.   □

Now, with the theory of analytic continuation for generalised hypergeometric functions, we can get access to the Mahler measure $\mu_q(k)$ itself.

Proposition 5.2 *For $q = 2, 3, \ldots$ and real $k > 0$, we have*

$$\mu_q(k) = \mathrm{Re}\left(\log k - \frac{2^{q-1}}{k^2}\,_{q+2}F_{q+1}\!\left(\begin{matrix} 1, 1, \frac{3}{2}, \ldots, \frac{3}{2} \\ 2, 2, \ldots, 2 \end{matrix}\,\middle|\, \frac{4^q}{k^2}\right)\right),$$

with $\frac{3}{2}$ repeated q times on the top and 2 repeated $q + 1$ times on the bottom.

Proof   Integrating the series expansion

$$\frac{1}{\pi^{q-1}k}\,F_q\!\left(\frac{4^q}{k^2}\right) = \sum_{n=0}^{\infty} \frac{(\frac{1}{2})_n^q}{n!^q}\,\frac{4^{qn}}{k^{2n+1}}$$

with respect to $k \in \mathbb{C} \setminus (-\infty, 0]$ in the domain $|k| > 2^q$ produces the expression

$$\log k - \frac{1}{2}\sum_{n=1}^{\infty} \frac{(\frac{1}{2})_n^q}{n!^q\,n}\,\frac{4^{qn}}{k^{2n}} = \log k - \frac{2^{q-1}}{k^2}\sum_{n=0}^{\infty} \frac{(\frac{3}{2})_n^q}{(2)_n^q\,(n+1)}\,\frac{4^{qn}}{k^{2n}}$$

$$= \log k - \frac{2^{q-1}}{k^2}\,_{q+2}F_{q+1}\!\left(\begin{matrix} 1, 1, \frac{3}{2}, \ldots, \frac{3}{2} \\ 2, 2, \ldots, 2 \end{matrix}\,\middle|\, \frac{4^q}{k^2}\right).$$

The hypergeometric function admits the analytic continuation to $k \in \mathbb{C} \setminus (-\infty, 2^q]$, with the same real parts along the two banks of the cut $(-\infty, 2^q]$: this follows from the real-valuedness of the function for real $k \in (2^q, \infty)$ and the Schwarz reflection principle.

Finally, we notice that the result of integration is continuous at $k = 4$ for $q = 2$, since Abel's theorem is applied to the resulting $_4F_3(16/k^2)$ series: the latter converges as $k \to 4^+$.   □

Exercise 5.3   Using the explicit formula for $\mu_1(k) = \lambda^+(k)$ from Section 1.4 or otherwise, verify that Proposition 5.2 is valid for $q = 1$ as well.

It is a common feature of all families of Mahler measures we consider below that the derivative with respect to their parameter k can be given by a convergent series, for sufficiently large $|k|$, that satisfies a homogeneous linear differential equation. This equation is not necessarily of hypergeometric type but can be reduced to a hypergeometric form with some further machinery.

5.2 Boyd's list and other instances

One significance of the Mahler measure would have been lost without its mysterious cast in the framework of Beilinson's conjectures, about the connection between regulator maps and values of L-functions. This was originally set up in Deninger's paper [68], followed by the works of Boyd [39] and Rodriguez Villegas [165]. The observation, in brief, is that for certain families $P_k(x, y)$ of polynomials their logarithmic Mahler measures m$(P_k(x, y))$ are rationally proportional to the L-values attached to the elliptic curves $P_k(x, y) = 0$. One particular example — that started the whole business in this direction — is Deninger's prediction [68]

$$m(x + 1/x + y + 1/y + 1) = rL'(E, 0)$$

for some rational r and $E : x + 1/x + y + 1/y + 1 = 0$ the elliptic curve of conductor 15. An extensive computation by Boyd [39] for this and many other examples offered $r = 1$; this was proved some 15 years later in the joint paper [170] by Rogers, a former student of Boyd, and Zudilin.

Postponing this remarkable intrigue till a better moment (namely, to Chapter 9; also check Section 8.4 for a more detailed historical account of the related development), here we try to highlight some particular players in the story — the Mahler measures of polynomials whose zero loci define elliptic and, in some rare situations, hyperelliptic curves. One such example,

$$\mu(k) = m(x + x^{-1} + y + y^{-1} + k) = m((x + x^{-1})(y + y^{-1}) + k) = \mu_2(k),$$

has already been discussed in Sections 3.5 and 5.1, while other instances [39, 165, 168] are the Mahler measures of

$$x^3 + y^3 + 1 - kxy, \quad (1+x)(1+y)(x+y) - kxy \quad \text{and} \quad (1+x)(1+y)(1+x+y) - kxy;$$

k serves as the parameter of families. In [23], the 'elliptic' Mahler measure of $y^3 - y + x^3 - x + kxy$ is related to the two 'hyperelliptic' Mahler measures

$$(x^2 + x + 1)y^2 + kx(x + 1)y + x(x^2 + x + 1),$$
$$(x^2 + x + 1)y^2 + (x^4 + kx^3 + (2k - 4)x^2 + kx + 1)y + x^2(x^2 + x + 1),$$

both taken from [39].

Exercise 5.4 (Rogers, Samart) Verify that the curve $kxy + (x + 1)^2(y^2 + 1) = 0$ is elliptic and defined over $\mathbb{Q}$ when $k^2 \in \mathbb{Z}$. Show that

$$w(k) = m(kxy + (x + 1)^2(y^2 + 1)) = m((x + 1/x)^2(y + 1/y) + k)$$

$$= \text{Re}\left(\log k - \frac{6}{k^2} \, _4F_3\left(\begin{matrix} 1, 1, \frac{5}{4}, \frac{7}{4} \\ 2, 2, 2 \end{matrix} \,\middle|\, \frac{64}{k^2}\right)\right)$$

and, for sufficiently large k,

$$w(k) = \mathrm{m}(e^{\pi i/4}\sqrt{k} + i(x + 1/x) + y + 1/y) = 2\,\mathrm{m}(y + x/y + 1/xy - \sqrt{k}).$$

Hint Verify that the curve $kxy + (x + 1)^2(y^2 + 1) = 0$ admits a Weierstrass form

$$Y^2 = X^3 - \frac{16k^6(k^2 - 48)}{3} X + \frac{128k^{10}(k^2 - 72)}{27}. \qquad \square$$

As functions of a parameter, the Mahler measures happen to satisfy numerous functional equations [119, 169] that may be thought of as generalisations of $\log|xy| = \log|x| + \log|y|$. A notable example here is

$$2\mu\left(2k + \frac{2}{k}\right) = \mu(4k^2) + \mu\left(\frac{4}{k^2}\right)$$

from [116].

5.3 Three-variate Mahler measures

There are several ways to adapt the two-variate families from the previous sections to three variables. Bertin's original paper [20] discussed two such examples, among which we single out

$$g_1(k) = \mathrm{m}(x + x^{-1} + y + y^{-1} + z + z^{-1} + k),$$

as it gives a generalisation of the measures $\lambda^+(k)$ from Section 1.4 and $\mu(k)$ which is different from $\mu_3(k)$ treated in Section 5.1. If we treat the family $g_1(k)$ using our previous strategy, then the result is as follows.

Exercise 5.5 Show that

$$g_1(k) = \mathrm{Re}\left(\log k - \sum_{n=1}^{\infty} \frac{k^{-2n}}{2n}\binom{2n}{n}\sum_{\ell=0}^{n}\binom{2\ell}{\ell}\binom{n}{\ell}^2\right)$$

for $k > 6$.

Later, in [167], Rogers complements Bertin's families with $\mu_3(k)$ and

$$f_3(k) = \mathrm{m}((x + x^{-1})^2(y + y^{-1})^2(1 + z)^3 - z^2 k),$$
$$f_4(k) = \mathrm{m}(x^4 + y^4 + z^4 + 1 + xyz\,k)$$

(our notation differs from that used in [167]), the last entry briefly mentioned by Rodriguez Villegas in the finale of [165]. An analysis of the latter two families, similar to the one in Section 5.1, gives hypergeometric formulae for the derivatives as well as for the Mahler measures themselves.

Exercise 5.6 (Rogers [167]) For real $k > 0$, show that

$$f_3(k) = \mathrm{Re}\!\left(\log k - \frac{12}{k}\, {}_5F_4\!\left(\begin{matrix} 1,\, 1,\, \frac{3}{2},\, \frac{4}{3},\, \frac{5}{3} \\ 2,\, 2,\, 2,\, 2 \end{matrix} \,\middle|\, \frac{108}{k}\right)\right),$$

$$f_4(k) = \mathrm{Re}\!\left(\log k - \frac{6}{k^4}\, {}_5F_4\!\left(\begin{matrix} 1,\, 1,\, \frac{3}{2},\, \frac{5}{4},\, \frac{7}{4} \\ 2,\, 2,\, 2,\, 2 \end{matrix} \,\middle|\, \frac{256}{k^4}\right)\right).$$

Chapter notes

Rogers uses in [167] the hypergeometric evaluations of three-variate Mahler measures as a source for producing new Ramanujan-type formulae for $1/\pi$.

More details on the latest advances in the resolution of Boyd's two-variate Mahler measures conjectures [39] are highlighted later, in Section 8.4. Here we only mention references Brunault [48, 51], Mellit [137] and Rogers and Zudilin [169, 170, 223]. The three-variate examples pioneered in [20] have received new life in the work of Samart [172, 173] and more recently of Brunault and Neururer [53], built on the work [52].

It is worth mentioning that the multivariate Mahler measure is naturally linked with asymptotic problems in analysis and statistical mechanics, and those offer different methods to treat the measure. One such instance is Szegő's first limit theorem [206], which describes the asymptotics of the 'block' Toeplitz determinant

$$T_N(P) = \det(a_{n-m})_{n,m\in\{1,\dots,N\}^k}$$

attached to a non-zero Laurent polynomial

$$P(x_1, \dots, x_k) = \sum_{n\in\mathbb{Z}^k} a_n x_1^{n_1} \cdots x_k^{n_k}$$

taking *real* values on the torus. Namely, the theorem asserts that

$$\lim_{N\to\infty} \frac{\log |T_N(P)|}{N^d} = m(P);$$

see [211] for details. This can be linked to the fact that the free energy of the Ising model can also be given as a Mahler measure. Indeed, the Hamiltonian of the Ising model (without external magnetic field) is given by

$$H(s) = -J \sum_{i\sim j} s_i s_j,$$

where $s_i \in \{\pm 1\}$ are spins and the sum is over all nearest neighbours $i \sim j$ within a given lattice. The associated partition function is $Z = \sum_s e^{-\beta H(s)}$, where the parameter β is the inverse temperature of the system. Finally, the free energy is obtained as a thermodynamic limit (the number of particles goes to ∞)

$$F_\beta = \lim_{n\to\infty} \frac{\log Z}{-\beta n}.$$

(For simplicity one may fix β to be 1.) The free energy can then be identified as the logarithmic Mahler measure of certain explicit polynomials [211]; for example, in the case of the two-dimensional Ising model on the square lattice $F = -\frac{1}{2} \text{m}(P)$, where

$$
\begin{aligned}
P(x,y) = P_T(x,y) &= 4\left(\frac{1+T^2}{1-T^2}\right)^2 - \frac{4T}{1-T^2}(x+1/x+y+1/y) \\
&= 4\cosh^2(2J) - 4\sinh^2(2J)(x+1/x+y+1/y), \\
T = \tanh J &= \frac{e^J - e^{-J}}{e^J + e^{-J}}.
\end{aligned}
$$

The calculation was carried out by Onsager in his famous 1944 tour-de-force solution of the model [149]; the connection of this computation (and related ones for other planar models) to Szegő's theorem, hence to a Mahler measure, can be implicitly observed in the Fan–Wu general technology [85]. The polynomial $P(x,y)$ possesses symmetries of the underlying lattice ($x \mapsto x^{-1}$ and $y \mapsto y^{-1}$), though its explicit form (and even the existence of a polynomial whose Mahler measure governs the free energy!) is somewhat mysterious.

Additional exercises

The next three exercises primarily deal with the so-called Euler–Gauss hypergeometric function (sometimes, referred to as the Gauss hypergeometric function)

$$F(a,b;c;z) = {}_2F_1\!\left(\begin{matrix} a, & b \\ & c \end{matrix}\,\middle|\, z\right),$$

which is just the particular ${}_2F_1$ case of the generalised hypergeometric function ${}_qF_{q-1}$ defined in Section 5.1. However, there are generalisations of those exercises to ${}_qF_{q-1}$ cases, of course, to some extent.

Exercise 5.7 (Contiguous relations) (a) Show that

$$\frac{\mathrm{d}}{\mathrm{d}z}F(a,b;c;z) = \frac{ab}{c}F(a+1,b+1;c+1;z),$$

$$F(a,b+1;c;z) - F(a,b;c;z) = \frac{az}{c}F(a+1,b+1;c+1;z).$$

(b) Two hypergeometric functions are said to be contiguous if their corresponding parameters a, b and c coincide, except for one pair of parameters in which they differ by 1; the parameter z is the same. Thus, $F(a, b; c; z)$ is contiguous to the six functions

$$F(a \pm 1, b; c; z), \quad F(a, b \pm 1; c; z) \quad \text{and} \quad F(a, b; c \pm 1; z).$$

Prove that any three of these functions are connected by a linear relation in z. Such relations are called contiguous.

(c) Show that any series $F(a + k_1, b + k_2; c + k_3; z)$, where k_1, k_2 and k_3 are integers, can be expressed as a $\mathbb{C}(z)$-linear combination of $F(a, b; c; z)$ and (any) one of its contiguous functions.

Comment   (b) The complete list of such three-term relations is reproduced in [194, Section 1.4].   □

Exercise 5.8 (Evaluation of the Euler–Gauss hypergeometric function)   (a) If $\operatorname{Re} c > \operatorname{Re} b > 0$ and $|z| < 1$, prove the Euler–Pochhammer integral representation

$$F(a, b; c; z) = \frac{\Gamma(c)}{\Gamma(b)\Gamma(c - b)} \int_0^1 t^{b-1}(1 - t)^{c-b-1}(1 - zt)^{-a} \, dt.$$

Show that the integral on the right-hand side converges for any $z \notin [1, +\infty)$, hence the formula provides us with the analytic continuation of hypergeometric series from the disc $|z| < 1$ to the domain $\mathbb{C} \setminus [1, +\infty)$.

(b) By comparing the two sides in the series expansion, verify that

$$F\left(a, b; c; -\frac{z}{1 - z}\right) = (1 - z)^a F(a, c - b; c; z)$$

if $|z| < 1$ and $\operatorname{Re} z < \frac{1}{2}$.

(c) Give Gauss's evaluation

$$F(a, b; c; 1) = \frac{\Gamma(c)\Gamma(c - a - b)}{\Gamma(c - a)\Gamma(c - b)}$$

provided that $\operatorname{Re} c > \operatorname{Re}(a + b)$.

(d) Give closed forms for

$$F(a, b; 1 + b - a; -1) \quad \text{and} \quad F(a, 1 - a; c; \tfrac{1}{2}).$$

(e) Prove that, for k a non-negative integer,

$$F\left(-\frac{k}{2}, \frac{1}{2} - \frac{k}{2}; \frac{3}{2} + k; -\frac{1}{3}\right) = \left(\frac{8}{9}\right)^k \frac{\Gamma(3/2 + k)/\Gamma(3/2)}{\Gamma(4/3 + k)/\Gamma(4/3)}.$$

Use Carlson's theorem [9, Section 5.3] to conclude that the identity is true for *any* k with $\operatorname{Re} k > -\frac{3}{2}$.

(f) For *a* complex with Re $a > -\frac{3}{4}$, compute

$$F\left(-a, 1 + 3a; \frac{3}{2} + 2a; \frac{1}{4}\right).$$

In particular, give a closed form for $F(\frac{1}{4}, \frac{1}{4}; 1; \frac{1}{4})$.

Exercise 5.9 (Solutions of the hypergeometric equation) (a) Show that the function

$$z^{1-c}F(a + 1 - c, b + 1 - c; 2 - c; z), \quad \text{where } |z| < 1, \ z \notin (-1, 0],$$

satisfies the hypergeometric differential equation

$$\left(\left(z\frac{d}{dz} + a\right)\left(z\frac{d}{dz} + b\right) - \left(z\frac{d}{dz}\right)\left(z\frac{d}{dz} + c - 1\right)\right)y = 0. \tag{5.2}$$

(Note that $F(a, b; c; z)$ satisfies the same equation.)

(b) If $c \neq 1$, then $F(a, b; c; z)$ and $z^{1-c}F(a + 1 - c, b + 1 - c; 2 - c; z)$ are two solutions of the *linear* hypergeometric differential equation (5.2), which are linearly independent over $\mathbb{C}$. If $c = 1$, then $F(a, b; 1; z)$ and

$$\sum_{n=0}^{\infty} \frac{(a)_n (b)_n}{n!^2} z^n \left(\log z - \sum_{k=0}^{n-1}\left(\frac{1}{a + k} + \frac{1}{b + k} - \frac{2}{1 + k}\right)\right)$$

are linearly independent solutions of the corresponding differential equation (5.2).

(c) If Re$(c - a - b) > 0$ and Re $c > 0$, show that

$$F(a, b; c; z) = \frac{\Gamma(c)\Gamma(c - a - b)}{\Gamma(c - a)\Gamma(c - b)} F(a, b; a + b - c + 1; 1 - z)$$

$$+ \frac{\Gamma(c)\Gamma(a + b - c)}{\Gamma(a)\Gamma(b)}(1 - z)^{c-a-b}F(c - a, c - b; 1 + c - a - b; 1 - z).$$

(d) For $z \in \mathbb{C} \setminus [0, +\infty)$ and suitable restrictions on a, b and c, prove that

$$F(a, b, c; z) = \frac{\Gamma(a)\Gamma(b - a)\Gamma(c)}{\Gamma(b)\Gamma(c - a)}(-z)^{-a}F(a, 1 + a - c, 1 + a - b; 1/z)$$

$$+ \frac{\Gamma(a - b)\Gamma(c)}{\Gamma(a)\Gamma(c - b)}(-z)^{-b}F(b, 1 + b - c, 1 + b - a; 1/z),$$

where all hypergeometric functions are the analytic continuations of the hypergeometric series (for example, from Exercise 5.8(a)) and $(-z)^c = \exp(c\log(-z))$ for the principal branch of the logarithm.

Hints (b) When $c = 1$, you can appeal to *Frobenius's method*, which uses the analytic dependence of the solutions on the parameter c. Since

$$\frac{z^{1-c}F(a + 1 - c, b + 1 - c, 2 - c; z) - F(a, b, c; z)}{1 - c}$$

solves equation (5.2) for $c \neq 1$, the limit as $c \to 1$ produces a solution for $c = 1$. Computing the limit using, for example, L'Hôpital's rule leads to the indicated formula involving $\log z$.

(c) First demonstrate that

$$F(a, b; a + b - c + 1; 1 - z) \quad \text{and} \quad (1 - z)^{c-a-b} F(c - a, c - b; 1 + c - a - b; 1 - z)$$

are two solutions of (5.2) and then find the connecting coefficients in the domain $|z| < 1$, $|1 - z| < 1$ by considering the limits as $z \to 0^+$ and $z \to 1^-$. Alternatively, use Exercise 5.8(a). □

Exercise 5.10 (Clausen's formula [60]) Prove the formula

$$_2F_1\!\left(\begin{matrix} a, b \\ a + b + \frac{1}{2} \end{matrix} \middle|\, z\right)^2 = {_3F_2}\!\left(\begin{matrix} 2a, 2b, a + b \\ a + b + \frac{1}{2}, 2a + 2b \end{matrix} \middle|\, z\right), \quad |z| < 1$$

and its variant

$$_2F_1\!\left(\begin{matrix} 2a, 2b \\ a + b + \frac{1}{2} \end{matrix} \middle|\, z\right)^2 = {_3F_2}\!\left(\begin{matrix} 2a, 2b, a + b \\ a + b + \frac{1}{2}, 2a + 2b \end{matrix} \middle|\, 4z(1 - z)\right),$$

valid in the left half of the lemniscate $|z(1 - z)| = \frac{1}{4}$.

Exercise 5.11 (Ramanujan's formula for $1/\pi$ [97]) (a) Prove that

$$\lim_{z \to 1^-} \sqrt{1 - z} \sum_{n=0}^{\infty} \frac{(a)_n (\frac{1}{2})_n (1 - a)_n}{n!^3} n z^n = \frac{\sin \pi a}{\pi}.$$

(b) Verify the following identity [167]:

$$_3F_2\!\left(\begin{matrix} \frac{1}{4}, \frac{1}{2}, \frac{3}{4} \\ 1, 1 \end{matrix} \middle|\, \frac{256x}{(1 + 3x)^4}\right) = \frac{1 + 3x}{1 + x/3} \, {_3F_2}\!\left(\begin{matrix} \frac{1}{4}, \frac{1}{2}, \frac{3}{4} \\ 1, 1 \end{matrix} \middle|\, \frac{256x^3}{9(3 + x)^4}\right).$$

(c) Show that

$$\sum_{n=0}^{\infty} \frac{(\frac{1}{4})_n (\frac{1}{2})_n (\frac{3}{4})_n}{n!^3} (40n + 3) \frac{1}{74^n} = \frac{49\sqrt{3}}{9\pi}.$$

Hints (a) Use

$$\lim_{n \to \infty} \frac{(a)_n (1 - a)_n n}{n!^2} = \frac{1}{\Gamma(a)\Gamma(1 - a)} = \frac{\sin \pi a}{\pi}$$

and the Stolz–Cesàro theorem.

(b) One way to proceed is by dividing the identity by $1 + 3x$ and verifying that both sides satisfy the same (linear) differential equation of order 3.

(c) Apply the differential operator $x \frac{d}{dx}$ to both sides of the identity in part (b) and pass to the limit as $x \to (1/9)^-$ using part (a). □

6

Random walk

The (multivariate) Mahler measure is the subject of (multiple!) generalisations — see, for example, [115, 153]. One variation is the zeta Mahler measure [4]

$$Z(P; s) = \int \cdots \int_{[0,1]^k} |P(e^{2\pi i t_1}, \ldots, e^{2\pi i t_k})|^s \, dt_1 \cdots dt_k,$$

of a non-zero Laurent polynomial $P(x_1, \ldots, x_k)$, so that

$$m(P) = \frac{dZ(P; s)}{ds}\bigg|_{s=0}.$$

In this chapter we examine polynomials P having a particular shape and related to random walks in the plane.

6.1 Density of a random walk

A k-step uniform random walk is a planar walk that starts at the origin and consists of k steps of length 1 each taken in a uniformly random direction. Let X_k denote the distance to the origin after these k steps. The sth moments $W_k(s)$ of X_k can be computed [34] via the formula

$$W_k(s) = \int \cdots \int_{[0,1]^k} |e^{2\pi i t_1} + \cdots + e^{2\pi i t_k}|^s \, dt_1 \cdots dt_k$$

$$= \int \cdots \int_{[0,1]^{k-1}} |1 + e^{2\pi i t_1} + \cdots + e^{2\pi i t_{k-1}}|^s \, dt_1 \cdots dt_{k-1}.$$

These can be recognised as the zeta Mahler measure of the linear polynomial $1 + x_1 + \cdots + x_{k-1}$, and are related to the (probability) density function $p_k(x)$ of

X_k via

$$W_k(s) = \int_0^\infty x^s p_k(x)\,dx = \int_0^k x^s p_k(x)\,dx.$$

That is, $p_k(x)$ can then be obtained as the inverse Mellin transform of $W_k(s-1)$.

More generally, we can consider (sufficiently nice, independent) random variables X_1 and X_2 on the ray $[0,\infty)$ with probability densities $p_1(x)$ and $p_2(x)$, respectively. If θ_1 and θ_2 are uniformly distributed on $[0,1]$, then $X = e^{2\pi i \theta_1} X_1 + e^{2\pi i \theta_2} X_2$ describes a two-step random walk in the plane with the first step of length X_1 and the second step of length X_2. An application of the cosine rule shows that the sth moment of $|X|$ is

$$W(s) = E(|X|^s) = \int_0^\infty \int_0^\infty g_s(x,y) p_1(x) p_2(y)\,dx\,dy,$$

where

$$g_s(x,y) = \frac{1}{\pi} \int_0^\pi (x^2 + y^2 + 2xy\cos\theta)^{s/2}\,d\theta.$$

Lemma 6.1 *We have*

$$W'(0) = E(\log|X|) = \int_0^\infty \int_0^\infty p_1(x) p_2(y) \max\{\log x, \log y\}\,dy\,dx.$$

Proof Observe that

$$\frac{dg_s(x,y)}{ds}\bigg|_{s=0} = \frac{1}{\pi} \int_0^\pi \log\sqrt{x^2 + y^2 + 2xy\cos\theta}\,d\theta = \max\{\log|x|, \log|y|\}$$

(compare with Exercise 1.9), and the result follows. □

Alternative equivalent expressions of Lemma 6.1 include

$$E(\log|X|) = \int_0^\infty \int_0^x p_1(x) p_2(y) \log x\,dy\,dx + \int_0^\infty \int_x^\infty p_1(x) p_2(y) \log y\,dy\,dx$$

$$= E(\log X_1) + \int_0^\infty \int_x^\infty p_1(x) p_2(y)(\log y - \log x)\,dy\,dx$$

$$= E(\log X_2) + \int_0^\infty \int_0^x p_1(x) p_2(y)(\log x - \log y)\,dy\,dx. \tag{6.1}$$

This can be regarded as a generalisation of Jensen's formula (Proposition 1.4).

6.2 Linear Mahler measures

We can summarise the characteristics of a uniform random walk from the previous section as

$$W_k(s) = Z(x_1 + \cdots + x_k; s) = Z(1 + x_1 + \cdots + x_{k-1}; s)$$

and

$$W'_k(0) = m(x_1 + \cdots + x_k) = m(1 + x_1 + \cdots + x_{k-1}) = \int_0^k p_k(x) \log x \, dx, \quad (6.2)$$

where the derivative is with respect to s. The latter Mahler measures are known as linear Mahler measures, and in Section 3.3 we discussed their evaluation for $k = 3$ and 4:

$$W'_3(0) = L'(\chi_{-3}, -1) = \frac{3\sqrt{3}}{4\pi} L(\chi_{-3}, 2), \quad W'_4(0) = -14\zeta'(-2) = \frac{7\zeta(3)}{2\pi^2},$$

due to Smyth. We can complement these formulae with the trivial instance $W'_2(0) = 0$, and with highly advanced, conjectural evaluations due to Rodriguez Villegas [42]:

$$W'_5(0) \overset{?}{=} -L'(f_{15}^{(3)}, -1) = 6\left(\frac{\sqrt{15}}{2\pi}\right)^5 L(f_{15}^{(3)}, 4),$$

$$W'_6(0) \overset{?}{=} -8L'(f_6^{(4)}, -1) = 3\left(\frac{\sqrt{6}}{\pi}\right)^6 L(f_6^{(4)}, 5),$$

where

$$f_{15}^{(3)}(\tau) = \eta_1^3 \eta_{15}^3 + \eta_3^3 \eta_5^3 \quad \text{and} \quad f_6^{(4)}(\tau) = \eta_1^2 \eta_2^2 \eta_3^2 \eta_6^2$$

are cusp eigenforms of weights 3 and 4, respectively. Here and in what follows we set $\eta_m = \eta(m\tau)$ with Dedekind's eta function

$$\eta(\tau) = q^{1/24} \prod_{m=1}^{\infty} (1 - q^m) = \sum_{n=-\infty}^{\infty} (-1)^n q^{(6n+1)^2/24}, \quad \text{where } q = e^{2\pi i \tau}, \quad (6.3)$$

serving as a principal constructor of modular forms and functions.

Importantly enough, the densities $p_k(x)$ of short uniform walks, namely for $k \leq 4$, are explicitly computed in terms of hypergeometric functions — see [34], especially Theorem 4.9 there for the case $k = 4$. Quite straightforwardly, we have $p_1(x) = \delta(x - 1)$ (the Dirac delta function) and $p_2(x) = 2/(\pi\sqrt{4 - x^2})$ for $0 < x < 2$, followed by

$$p_3(x) = \frac{2\sqrt{3}x}{\pi(3 + x^2)} \, {}_2F_1\left(\begin{matrix} \frac{1}{3}, \frac{2}{3} \\ 1 \end{matrix} \middle| \frac{x^2(9 - x^2)^2}{(3 + x^2)^3}\right)$$

for $0 < x < 3$, and

$$p_4(x) = \frac{2\sqrt{16 - x^2}}{\pi^2 x} \operatorname{Re} {}_3F_2\left(\begin{matrix} \frac{1}{2}, \frac{1}{2}, \frac{1}{2} \\ \frac{5}{6}, \frac{7}{6} \end{matrix} \middle| \frac{(16 - x^2)^3}{108x^4}\right)$$

for $0 < x < 4$.

Exercise 6.1 ([203]) Let k, m be integers such that $k > m > 0$. Show that

$$W_k'(0) = W_m'(0) + \int_0^{k-m} p_{k-m}(x)\left(\int_0^x p_m(y)(\log x - \log y)\,dy\right)dx.$$

Use the formula and known densities to write $W_5'(0)$ and $W_6'(0)$ as single integrals of elementary and hypergeometric functions.

Hint Decompose a k-step random walk into two walks with $k - m$ and m steps, and apply Lemma 6.1 in the form (6.1). □

The densities $p_3(x)$ and $p_4(x)$ can also be expressed by means of modular functions and forms [34, 203]. The related modular parametrisation of $p_3(x)$ is given by

$$x = x(\tau) = 3\frac{\eta(\tau)^2\eta(6\tau)^4}{\eta(2\tau)^4\eta(3\tau)^2} : (i\infty, 0) \to (0, 3),$$

so that

$$p_3(x) = \frac{2\sqrt{3}}{\pi}\frac{\eta(2\tau)^2\eta(6\tau)^2}{\eta(\tau)\eta(3\tau)}.$$

For the density $p_4(x)$ we have

$$p_4(x(\tau)) = -\operatorname{Re}\left(\frac{2i(1 + 6\tau + 12\tau^2)}{\pi}p(\tau)\right),$$

where

$$p(\tau) = \frac{\eta(2\tau)^4\eta(6\tau)^4}{\eta(\tau)\eta(3\tau)\eta(4\tau)\eta(12\tau)} \quad \text{and} \quad x(\tau) = \left(\frac{2\eta(\tau)\eta(3\tau)\eta(4\tau)\eta(12\tau)}{\eta(2\tau)^2\eta(6\tau)^2}\right)^3.$$

The path for τ along the imaginary axis from 0 to $i/(2\sqrt{3})$ (or from $i\infty$ to $i/(2\sqrt{3})$) corresponds to x ranging from 0 to 2, while the path from $i/(2\sqrt{3})$ to $-1/4 + i/(4\sqrt{3})$ along the arc centred at 0 corresponds to the real range $(2, 4)$ for x. (The arc admits the parametrisation $\tau = e^{\pi i\theta}/(2\sqrt{3})$, $1/2 < \theta < 5/6$.)

The next exercise (after Ramanujan) strikingly resembles the evaluation of $W_3'(0)$ and is related to the modular parametrisation of the density $p_3(x)$, though a direct link between it and linear Mahler measure remains missing.

Exercise 6.2 ([18]) Show that

$$\int_0^1 \frac{1}{9}\left(1 - \frac{\eta(\tau)^9}{\eta(3\tau)^3}\right)\frac{dq}{q} = L'(\chi_{-3}, -1).$$

6.3 Random walk variations

As shown in [203], the 'random walk' technique of the previous sections can be successfully applied in somewhat 'non-uniform' settings to produce hypergeometric expressions for the Mahler measures in the following conjectural evaluations:

$$m(1 + x + y + xy + z) \overset{?}{=} -2L'(f_{15}, -1) = \frac{15^2}{4\pi^4}L(f_{15}, 3) = 0.4839979734\ldots,$$

$$m((1 + x)^2 + y + z) \overset{?}{=} -L'(f_{24}, -1) = \frac{72}{\pi^4}L(f_{24}, 3) = 0.7025655062\ldots,$$

$$m(1 + x + y - xy + z) \overset{?}{=} -\frac{5}{4}L'(f_{21}, -1) = \frac{5 \cdot 21^2}{32\pi^4}L(f_{21}, 3) = 0.6242499823\ldots,$$

the first attributed to Boyd and the second from [42]. Here $f_{15}(\tau) = \eta_1\eta_3\eta_5\eta_{15}$, $f_{24}(\tau) = \eta_2\eta_4\eta_6\eta_{12}$ and

$$f_{21}(\tau) = \frac{1}{36}\frac{\eta_1^{12}\eta_6^2}{\eta_2^6\eta_3^4} + \frac{7}{36}\frac{\eta_7^{12}\eta_{42}^2}{\eta_{14}^6\eta_{21}^4} - \frac{2}{9}\frac{\eta_1^6\eta_6\eta_7^6\eta_{42}}{\eta_2^3\eta_3^2\eta_{14}^3\eta_{21}^2} + 4\frac{\eta_1^2\eta_6^2\eta_7^2\eta_{42}^2}{\eta_2\eta_3\eta_{14}\eta_{21}}$$

are cusp forms of weight 2 (and of levels 15, 24 and 21, respectively).

In analogy with the case of linear Mahler measures, we define

$$\tilde{W}(s) = \iiint_{[0,1]^3} |1 + e^{2\pi i t_1} + e^{2\pi i t_2} + e^{2\pi i t_3} + e^{2\pi i (t_2+t_3)}|^s \, dt_1 \, dt_2 \, dt_3$$

$$= Z(1 + x_1 + x_2 + x_3 + x_2x_3; s)$$

as the sth moment of a random 5-step walk for which the direction of the final step is completely determined by the previous two steps.

Theorem 6.2 ([203, Theorem 2]) *We have*

$$m(1 + x + y + xy + z) = -\frac{1}{2\pi}\int_0^1 {}_2F_1\left(\begin{matrix}\frac{1}{2}, \frac{1}{2} \\ 1\end{matrix} \, \middle| \, 1 - \frac{x^2}{16}\right)\log x \, dx.$$

Proof We use the notation x_1, x_2, x_3 for z, x, y, respectively. Define a related density $\hat{p}(x)$ by

$$\int_0^4 x^s \hat{p}(x) \, dx = \hat{W}(s) = \iint_{[0,1]^2} |1 + e^{2\pi i t_2} + e^{2\pi i t_3} + e^{2\pi i (t_2+t_3)}|^s \, dt_2 \, dt_3$$

$$= W_2(s)^2 = \frac{\Gamma(1 + s)^2}{\Gamma(1 + s/2)^4}.$$

By an application of the Mellin transform calculus, we find that for $0 < x < 4$,

$$\hat{p}(x) = \frac{1}{2\pi} {}_2F_1\left(\begin{matrix}\frac{1}{2}, \frac{1}{2} \\ 1\end{matrix} \, \middle| \, 1 - \frac{x^2}{16}\right).$$

Then it follows from Lemma 6.1 that

$$\tilde{W}'(0) = \int_1^4 \hat{p}(x) \log x \, dx = -\int_0^1 \hat{p}(x) \log x \, dx,$$

where we use the evaluation

$$\int_0^4 \hat{p}(x) \log x \, dx = m(1 + x_2 + x_3 + x_2 x_3) = m(1 + x_2) + m(1 + x_3) = 0. \quad \square$$

Exercise 6.3 (Cf. Exercise 3.6) Extend the above computation to the general formula

$$m((1+x)(1+y) + kz) = \log k \int_0^k \hat{p}(x) \, dx + \int_k^4 \hat{p}(x) \log x \, dx$$

$$= \frac{1}{2\pi} \int_0^k {}_2F_1\left(\begin{matrix} \frac{1}{2}, \frac{1}{2} \\ 1 \end{matrix} \middle| 1 - \frac{x^2}{16}\right) \log \frac{k}{x} \, dx$$

for $0 < k \le 4$.

The left-hand side of the level-24 Mahler measure conjecture can be treated by a similar reduction, using that the densities for $(1 + x)^2$ and $y + z$ are $p_2(t^{1/2})/(2t^{1/2})$ on $[0,4]$ and $p_2(t)$ on $[0,2]$, respectively. The final result is the elegant formula

$$m((1+x)^2 + y + z) = \frac{2G}{\pi} + \frac{2}{\pi^2} \int_0^1 \arcsin(1-x) \arcsin x \, \frac{dx}{x}, \tag{6.4}$$

where G is Catalan's constant (3.3), and, with some further work, we can express the right-hand side hypergeometrically.

Theorem 6.3 ([203, Theorem 3]) *We have*

$$m((1+x)^2 + y + z) = \frac{8\Gamma(\frac{3}{4})^2}{\pi^{5/2}} \, {}_5F_4\left(\begin{matrix} \frac{1}{4}, \frac{1}{4}, \frac{1}{4}, \frac{3}{4}, \frac{3}{4} \\ \frac{1}{2}, \frac{5}{4}, \frac{5}{4}, \frac{5}{4} \end{matrix} \middle| \frac{1}{4}\right)$$

$$+ \frac{\Gamma(\frac{1}{4})^2}{54\pi^{5/2}} \, {}_5F_4\left(\begin{matrix} \frac{3}{4}, \frac{3}{4}, \frac{3}{4}, \frac{5}{4}, \frac{5}{4} \\ \frac{3}{2}, \frac{7}{4}, \frac{7}{4}, \frac{7}{4} \end{matrix} \middle| \frac{1}{4}\right).$$

Proof Notice that, for $0 < x < 1$,

$$\arcsin(1 - x) = \frac{\pi}{2} - \arccos(1 - x) = \frac{\pi}{2} - \sqrt{2x} \, {}_2F_1\left(\begin{matrix} \frac{1}{2}, \frac{1}{2} \\ \frac{3}{2} \end{matrix} \middle| \frac{x}{2}\right),$$

and that, for $n > -1/2$,

$$\int_0^1 x^{n-1/2} \arcsin x \, dx = \frac{\sqrt{\pi}}{2n+1}\left(\sqrt{\pi} - \frac{\Gamma(\frac{n}{2} + \frac{3}{4})}{\Gamma(\frac{n}{2} + \frac{5}{4})}\right).$$

Therefore,

$$\int_0^1 \arcsin(1-x)\arcsin x \,\frac{\mathrm{d}x}{x} = \frac{\pi}{2}\int_0^1 \arcsin x \,\frac{\mathrm{d}x}{x}$$

$$-\pi\sqrt{2}\sum_{n=0}^{\infty}\frac{(\frac{1}{2})_n^2}{n!\,(\frac{3}{2})_n(2n+1)}\frac{1}{2^n}$$

$$+\sqrt{2\pi}\sum_{n=0}^{\infty}\frac{(\frac{1}{2})_n^2\Gamma(\frac{n}{2}+\frac{3}{4})}{n!\,(\frac{3}{2})_n(2n+1)\Gamma(\frac{n}{2}+\frac{5}{4})}\frac{1}{2^n}.$$

From this and (6.4) we deduce

$$\mathrm{m}((1+x)^2+y+z) = \frac{2G}{\pi}+\frac{\log 2}{2}-\frac{2\sqrt{2}}{\pi}\,{}_3F_2\!\left(\begin{matrix}\frac{1}{2},\frac{1}{2},\frac{1}{2}\\\frac{3}{2},\frac{3}{2}\end{matrix}\middle|\frac{1}{2}\right)$$

$$+\frac{8\sqrt{2}\,\Gamma(\frac{3}{4})}{\pi^{3/2}\Gamma(\frac{1}{4})}\,{}_5F_4\!\left(\begin{matrix}\frac{1}{4},\frac{1}{4},\frac{1}{4},\frac{3}{4},\frac{3}{4}\\\frac{1}{2},\frac{5}{4},\frac{5}{4},\frac{5}{4}\end{matrix}\middle|\frac{1}{4}\right)+\frac{\sqrt{2}\,\Gamma(\frac{1}{4})}{54\pi^{3/2}\Gamma(\frac{3}{4})}\,{}_5F_4\!\left(\begin{matrix}\frac{3}{4},\frac{3}{4},\frac{3}{4},\frac{5}{4},\frac{5}{4}\\\frac{3}{2},\frac{7}{4},\frac{7}{4},\frac{7}{4}\end{matrix}\middle|\frac{1}{4}\right).$$

It remains to use

$$G+\frac{1}{4}\pi\log 2 = \sqrt{2}\,{}_3F_2\!\left(\begin{matrix}\frac{1}{2},\frac{1}{2},\frac{1}{2}\\\frac{3}{2},\frac{3}{2}\end{matrix}\middle|\frac{1}{2}\right)$$

(see [2, Entry 30]) and $\Gamma(\frac{1}{4})\Gamma(\frac{3}{4})=\pi\sqrt{2}$. □

Exercise 6.4 Show that the Mahler measure $\mathrm{m}(1+x+y-xy+z)$ is equal to

$$\frac{1}{2\pi}\int_1^{\sqrt{8}}{}_2F_1\!\left(\begin{matrix}\frac{1}{2},\frac{1}{2}\\1\end{matrix}\middle|\frac{t^2}{2}-\frac{t^4}{16}\right)t\log t\,\mathrm{d}t.$$

Use this formula to verify numerically that it is $-\frac{5}{4}L'(f_{21},-1)$ (to 100 places, say).

Hint Observe that $0\le|1+x+y-xy+z|\le\sqrt{8}$ on the torus $|x|=|y|=|z|=1$. As in the proof of Theorem 6.2 show that the related density $p(t)$ defined by

$$\int_0^{\sqrt{8}}t^s p(t)\,\mathrm{d}t = \iint_{[0,1]^2}|1+e^{2\pi i t_1}+e^{2\pi i t_2}-e^{2\pi i(t_1+t_2)}|^s\,\mathrm{d}t_1\,\mathrm{d}t_2$$

is equal to

$$p(t) = \frac{t}{2\pi}\,{}_2F_1\!\left(\begin{matrix}\frac{1}{2},\frac{1}{2}\\1\end{matrix}\middle|\frac{t^2}{2}-\frac{t^4}{16}\right).$$

Finally, numerical calculation of $L'(f_{21},-1)$ can be performed in PARI/GP [151] using lfun(ellinit("21a1"),-1,1). □

Chapter notes

It should be remarked that numerical evaluation of the linear Mahler measure (6.2) for an arbitrary n already represents a challenge. This was addressed by Bailey and Borwein in [8, Section 5] based on representation of $W_n'(0)$ through Bessel functions and use of novel integration techniques; the computation gives the quantities with 1000-digit accuracy for an impressive range of n.

In [188] the linear Mahler measure $m(1 + x_1 + x_2 + x_3 + x_4)$ is reduced to a linear combination of double L-values of certain meromorphic modular forms of weight 4. Apart from the evaluations (partly conjectural) of the linear Mahler measures $m(1 + x_1 + \cdots + x_{k-1})$ for $k \leq 6$, no similar formulae are known when $k \geq 7$. However, the story continues at a different level — see [47, 221] for details.

The conjectural L-value evaluation of $m(1 + x + y + xy - z)$ and Theorem 6.2 brings us to the expectation

$$\frac{1}{2\pi} \int_0^1 {}_2F_1\left(\begin{matrix} \frac{1}{2}, \frac{1}{2} \\ 1 \end{matrix} \middle| 1 - \frac{x^2}{16}\right) \log x \, dx \overset{?}{=} 2L'(f_{15}, -1). \tag{6.5}$$

This one highly resembles the evaluation

$$\frac{1}{2} \int_0^1 {}_2F_1\left(\begin{matrix} \frac{1}{2}, \frac{1}{2} \\ 1 \end{matrix} \middle| \frac{x^2}{16}\right) dx = \frac{1}{2} {}_3F_2\left(\begin{matrix} \frac{1}{2}, \frac{1}{2}, \frac{1}{2} \\ 1, \frac{3}{2} \end{matrix} \middle| \frac{1}{16}\right) = 2L'(f_{15}, 0) \tag{6.6}$$

established in [170]. Furthermore, Cohen (2018) observes another step on the ladder (6.6), (6.5):

$$\frac{6}{\pi^2} \int_0^1 {}_2F_1\left(\begin{matrix} \frac{1}{2}, \frac{1}{2} \\ 1 \end{matrix} \middle| \frac{x^2}{16}\right) \log^2 x \, dx \overset{?}{=} 2L'(f_{15}, -2) = \frac{3 \cdot 15^3}{8\pi^6} L(f_{15}, 4)$$

$$= 1.2165632526\ldots,$$

though not linked to a particular Mahler measure.

The expression in Theorem 6.3 is somewhat different from the one in Theorem 6.2, and resembles the hypergeometric evaluation of the L-value

$$-L'(f_{32}, -1) = \frac{128}{\pi^4} L(f_{32}, 3)$$

$$= \frac{\Gamma(\frac{1}{4})^2}{6\sqrt{2}\pi^{5/2}} {}_4F_3\left(\begin{matrix} 1, 1, 1, \frac{1}{2} \\ \frac{7}{4}, \frac{3}{2}, \frac{3}{2} \end{matrix} \middle| 1\right) + \frac{4\Gamma(\frac{3}{4})^2}{\sqrt{2}\pi^{5/2}} {}_4F_3\left(\begin{matrix} 1, 1, 1, \frac{1}{2} \\ \frac{5}{4}, \frac{3}{2}, \frac{3}{2} \end{matrix} \middle| 1\right)$$

$$+ \frac{\Gamma(\frac{1}{4})^2}{2\sqrt{2}\pi^{5/2}} {}_4F_3\left(\begin{matrix} 1, 1, 1, \frac{1}{2} \\ \frac{3}{4}, \frac{3}{2}, \frac{3}{2} \end{matrix} \middle| 1\right),$$

where $f_{32}(\tau) = \eta_4^2 \eta_8^2$ is a cusp form of level 32, obtained in [222, Theorem 3].

We finally complement the conjectures from Section 6.3 relating three-variate Mahler measures to L-values with the following list:

$$m((1 + x)(1 + y)(x + y) + z) = m((x + 1/x)(y + 1/y)(x/y + y/x) + z)$$
$$\overset{?}{=} -3L'(f_{14}, -1) = 0.6233530933\ldots,$$
$$m((1 + x)^2(1 + y) + z) = m((x + 1/x)^2(y + 1/y) + z)$$
$$\overset{?}{=} -\frac{3}{2}L'(f_{21}, -1) = 0.7490999787\ldots,$$
$$m((1 + x)^2 + (1 - x)(y + z)) \overset{?}{=} -2L'(f_{20}, -1) = 0.9497793314\ldots,$$

where $f_{14}(\tau) = \eta_1\eta_2\eta_7\eta_{14}$ and $f_{20}(\tau) = \eta_2^2\eta_{10}^2$, and also with the entry

$$m(1 + (1 + x + x^2)y + (1 - x + x^2)z)$$
$$\overset{?}{=} -\frac{1}{6}L'(f_{45}, -1) + L'(\chi_{-3}, -1) = 0.7981546024\ldots$$

from Boyd's lesser-known list in [41], where

$$f_{45}(\tau) = \frac{\eta_{15}^{10}}{\eta_5^3\eta_{45}^3} - \frac{\eta_1\eta_9\eta_{15}^6}{\eta_5^2\eta_{45}^2} - \frac{\eta_3^2\eta_5^2\eta_{45}^2}{\eta_1\eta_9}$$

is the cusp eigenform of weight 2 and level 45 [126].

Additional exercises

The following exercise generalises the result from Exercise 6.2.

Exercise 6.5 ([3]) Let α be real, $k \geq 2$ integer, and let a non-trivial character χ satisfy $\chi(-1) = (-1)^k$. Show that

$$q^\alpha \prod_{n=1}^{\infty}(1 - q^n)^{\chi(n)n^{k-2}} = \exp\left(-L'(\chi, 2 - k) - \int_q^1\left(\alpha - \sum_{m=1}^{\infty}\sum_{d\mid m}\chi(d)d^{k-1}t^m\right)\frac{dt}{t}\right)$$

is valid for real q in the interval $0 < q < 1$.

Exercise 6.6 ([207]) Let $f(\tau) = \sum_{n=1}^{\infty} c(n)e^{2\pi i n\tau}$ be a holomorphic cusp form and α a real number.

(a) Show that

$$q^\alpha \prod_{n=1}^{\infty}(1 - q^n)^{c(n)} = \exp\left(-\Lambda(f, 0) - \int_q^1\left(\alpha - \sum_{m=1}^{\infty}\sum_{d\mid m}dc(d)t^m\right)\frac{dt}{t}\right)$$

is valid for real q in the interval $0 < q < 1$. Here

$$\Lambda(f, 0) = \int_0^\infty f(it) \frac{\mathrm{d}t}{t}$$

is the completed L-function of the cusp form f at 0.

(b) Give a related result for the product

$$q^\alpha \prod_{n=1}^\infty (1 - q^n)^{n^k c(n)},$$

where $k \geq 1$ is an integer.

7

The regulator map for K_2 of curves

The aim of this chapter is to explain how the Mahler measure of a two-variable polynomial can be expressed in terms of a *regulator map*. We first give the definition of the Bloch–Beilinson regulator map for the K_2 group of a complex algebraic curve [13], and then explain the connection with the Mahler measure following Deninger [68].

7.1 Algebraic curves and their K_2 groups

Let $P(x,y) \in \mathbb{C}[x,y]$ be an irreducible polynomial, and let $C_P : P(x,y) = 0$ denote the zero locus of P in $\mathbb{C}^2$. Removing the singular points from C_P, we get a complex-analytic manifold C_P^{reg} of (complex) dimension 1, in other words, a *Riemann surface*. Explicitly, we have

$$C_P^{\text{reg}} = \left\{(x,y) \in \mathbb{C}^2 : P(x,y) = 0, \ \left(\frac{\partial P}{\partial x}, \frac{\partial P}{\partial y}\right)(x,y) \neq (0,0)\right\}. \tag{7.1}$$

More generally, every non-singular (or smooth) complex algebraic curve X (not necessarily planar) gives rise to a Riemann surface $X(\mathbb{C})$, the set of complex points of X. We thus have a natural mapping

$$\left\{\begin{matrix}\text{smooth complex} \\ \text{algebraic curves}\end{matrix}\right\} \longrightarrow \{\text{Riemann surfaces}\}.$$

Conversely, let C be a *compact* Riemann surface — for example, one may think of the Riemann sphere $\mathbb{P}^1(\mathbb{C}) = \mathbb{C} \cup \{\infty\}$ or of an elliptic curve $\mathbb{C}/\Lambda$, where Λ is a lattice in $\mathbb{C}$ (see Figure 7.1). By Riemann's existence theorem, there is a smooth complex algebraic curve X such that the Riemann surface $X(\mathbb{C})$ is isomorphic to C. The curve X is projective since C is compact. Moreover, X is

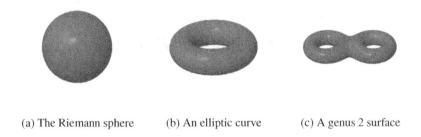

(a) The Riemann sphere   (b) An elliptic curve   (c) A genus 2 surface

Figure 7.1 Some compact Riemann surfaces

uniquely determined up to isomorphism. This way we get a correspondence

$$\left\{\begin{array}{c}\text{smooth projective}\\\text{complex algebraic curves}\end{array}\right\} \longleftrightarrow \{\text{compact Riemann surfaces}\} \qquad (7.2)$$

which is a bijection if we take the isomorphism classes on both sides. It is even an equivalence of categories: the algebraic maps $X \to X'$ on the left-hand side correspond exactly to the holomorphic maps $X(\mathbb{C}) \to X'(\mathbb{C})$ on the right-hand side.

The correspondence (7.2) can be extended to non-projective curves: if S is a finite set of points of C, then $C^* = C \setminus S$ is called a *compact Riemann surface with punctures*. Since every point of S can be seen as a closed point of X, the smooth algebraic curve $Y = X \setminus S$ satisfies $Y(\mathbb{C}) \cong C^*$. In this way we obtain a bijection between isomorphism classes of smooth complex algebraic curves on the one hand, and compact Riemann surfaces with punctures on the other hand. Note that every smooth algebraic curve has a unique smooth compactification, which can be obtained by adding finitely many closed points.

As an example, let Γ be a discrete subgroup of $\mathrm{SL}_2(\mathbb{R})$, acting on the upper half-plane $\mathcal{H} = \{z \in \mathbb{C} : \mathrm{Im}(z) > 0\}$ by Möbius transformations. Siegel [190] proved that the quotient $\Gamma \setminus \mathcal{H}$ is a compact Riemann surface with punctures if and only if Γ has finite covolume, that is, the area of a fundamental domain for the action of Γ on $\mathcal{H}$ is finite. In this case $\Gamma \setminus \mathcal{H}$ defines a smooth algebraic curve, called a *modular curve* and denoted by $Y(\Gamma)$.

We now come to the definition of the objects involved in the regulator map. In what follows, we fix a smooth connected complex algebraic curve Y, and denote by X its smooth compactification, so that $Y = X \setminus S$ for some finite set S of closed points of X. We will sometimes abusively use the same letter for both the algebraic curve and the associated Riemann surface.

The field of meromorphic functions on $X(\mathbb{C})$ is isomorphic to the *function field* $\mathbb{C}(X)$ of X, defined algebraically as the infinite union

$$\mathbb{C}(X) = \varinjlim_{\text{finite } T \subset X} O(X \setminus T),$$

where T runs through the finite sets of closed points of X, and $O(X \setminus T)$ denotes the ring of regular functions on $X \setminus T$. In particular, $O(Y)$ injects into $\mathbb{C}(X)$.

The Bloch–Beilinson regulator map associated to Y is a linear map

$$\mathrm{reg}_Y \colon K_2(Y) \to H^1(Y(\mathbb{C}), \mathbb{R}), \qquad (7.3)$$

where $K_2(Y)$ is Quillen's algebraic K-group associated to Y, and $H^1(Y(\mathbb{C}), \mathbb{R})$ is the de Rham cohomology of $Y(\mathbb{C})$. We will not define K_2 here (see [158]), but rather list some of its properties:

(1) For any (not necessarily commutative!) ring R, Milnor [140] constructs an abelian group $K_2(R)$ using relations between elementary matrices in the general linear group $\mathrm{GL}(R)$.

(2) For any commutative ring R, there is a canonical map [140, Section 8]

$$R^\times \otimes_{\mathbb{Z}} R^\times \to K_2(R). \qquad (7.4)$$

For any $x, y \in R^\times$, the *Milnor symbol* $\{x, y\}$ is the image of $x \otimes y$ in $K_2(R)$. We have the relations $\{y, x\} = -\{x, y\}$ and $\{x, -x\} = 0$. If x and $1 - x$ are both units of R, then $\{x, 1 - x\} = 0$ [140, Lemma 9.8].

(3) The group $K_2(R)$ is *functorial in R*: every morphism of rings $\varphi \colon R \to R'$ gives rise to a linear map $\varphi_* \colon K_2(R) \to K_2(R')$, and the symbol map (7.4) is compatible with morphisms of commutative rings.

(4) If $R = F$ is a *field*, then Matsumoto's theorem [140, Sections 11–12] asserts that the map (7.4) induces an isomorphism

$$K_2(F) \cong \frac{F^\times \otimes_{\mathbb{Z}} F^\times}{\langle x \otimes (1 - x) : x \in F \setminus \{0, 1\} \rangle_{\mathbb{Z}}}.$$

One should be careful that, for general R, the map (7.4) is not surjective. It is a difficult question to determine its kernel and cokernel. For example, it is an open problem in the case $R = O(Y)$ when the curve Y has genus ≥ 1.

(5) If the curve Y is affine (which amounts to saying that S is not empty), then Quillen's $K_2(Y)$ coincides with Milnor's $K_2(O(Y))$. Accordingly, every morphism of algebraic curves $\varphi \colon Y \to Y'$ induces a linear map $\varphi^* \colon K_2(Y') \to K_2(Y)$, called the *pull-back by φ*.

(6) By Quillen's localisation theorem [158], there is an exact *localisation se-quence*

$$\cdots \longrightarrow K_2(Y) \longrightarrow K_2(\mathbb{C}(X)) \xrightarrow{\partial} \bigoplus_{p \in Y} \mathbb{C}^\times \longrightarrow \cdots .$$

(7.5)

The first map comes from the functoriality of K_2, and the map $\partial = (\partial_p)_{p \in Y}$ is given by the *tame symbols* at the points of Y, namely,

$$\partial_p\{f, g\} = (-1)^{v_p(f)v_p(g)}\left(\frac{f^{v_p(g)}}{g^{v_p(f)}}\right)(p) \quad \text{for } f, g \in \mathbb{C}(X)^\times,$$

where v_p denotes the order of vanishing of a function at p. Note that the function $f^{v_p(g)}/g^{v_p(f)}$ has order of vanishing 0 at p, so its value at p is in $\mathbb{C}^\times$. (For Quillen's derivation of (7.5), see the exact sequence in the proof of [158, Theorem 5.4], with $i = 2$ and $p = 0$.)

The tame symbol ∂_p is a particular case of the construction given in the following exercise, noting that v_p is a discrete valuation on the field $\mathbb{C}(X)$.

Exercise 7.1 Let F be a field.

(a) Show that $\{x, -x\} = 0$ in $K_2(F)$ and then $\{y, x\} = -\{x, y\}$ for any $x, y \in F^\times$.
(b) Let v be a discrete valuation on F, with residue field k. Define the *tame symbol* ∂_v by

$$\partial_v : K_2(F) \to k^\times, \quad \{x, y\} \mapsto (-1)^{v(x)v(y)}\left(\frac{x^{v(y)}}{y^{v(x)}}\right)_v,$$

where $(\,\cdot\,)_v$ denotes the reduction modulo v. Show that ∂_v is a well defined surjective linear map.

Hint   (a) Use $\{x^{-1}, 1 - x^{-1}\} = 0$.   □

7.2 The regulator map

We now introduce a differential form, which is at the heart of the definition of the regulator.

Definition 7.1 Let $X/\mathbb{C}$ be a smooth connected projective curve. Let $f, g \in \mathbb{C}(X)^\times$. Let $S_{f,g}$ be the finite set of zeros and poles of f and g in X. Denote by $\eta(f, g)$ the following real analytic differential 1-form on $X \setminus S_{f,g}$:

$$\eta(f, g) = \log|f|\, d\arg(g) - \log|g|\, d\arg(f),$$

where $d\arg(h)$ is defined as the imaginary part of $d\log(h) = (dh)/h$.

Note that the map $(f, g) \in \mathbb{C}(X)^\times \times \mathbb{C}(X)^\times \mapsto \eta(f, g)$ is bilinear and anti-symmetric (here we view the multiplicative group $\mathbb{C}(X)^\times$ as a $\mathbb{Z}$-module). The following exercise gives an important property of $\eta(f, g)$.

Exercise 7.2 Show that the differential form $\eta(f, g)$ is closed.

This implies that the integral of $\eta(f, g)$ over a path γ contained in $X \setminus S_{f,g}$ depends only on the homotopy class of γ.

Though the definition of $\eta(f, g)$ is given for curves X, it still makes perfect sense when f and g are two meromorphic functions on a complex-analytic manifold of arbitrary dimension. However,

$$d\eta(f, g) = \frac{1}{2i}\left(\frac{df}{f} \wedge \frac{dg}{g} - \frac{d\overline{f}}{\overline{f}} \wedge \frac{d\overline{g}}{\overline{g}}\right) = \mathrm{Im}(d\log(f) \wedge d\log(g)),$$

so that $\eta(f, g)$ need not be closed in general.

Theorem 7.2 *For $f \in \mathbb{C}(X) \setminus \{0, 1\}$, the differential form $\eta(f, 1 - f)$ is exact, and one primitive is given by the function $D \circ f$, where $D: \mathbb{P}^1(\mathbb{C}) \to \mathbb{R}$ is the Bloch–Wigner dilogarithm (4.3).*

Proof The function D is a primitive of $\eta(z, 1 - z)$ on $\mathbb{P}^1(\mathbb{C}) \setminus \{0, 1, \infty\}$ by Exercise 4.4(c). Pulling back to X using $f: X \setminus S_{f,1-f} \to \mathbb{P}^1(\mathbb{C}) \setminus \{0, 1, \infty\}$, it follows that $\eta(f, 1 - f) = d(D \circ f)$. $\square$

In order to study the behaviour of $\eta(f, g)$ at the zeros and poles of f and g, we use the notion of residues.

Definition 7.3 Let $f, g \in \mathbb{C}(X)^\times$ and $p \in X$. The *residue* of $\eta(f, g)$ at p is

$$\mathrm{Res}_p(\eta(f, g)) = \int_{\gamma_p} \eta(f, g),$$

where γ_p is a sufficiently small loop around p (oriented counterclockwise), that is, γ_p is the boundary of a disc containing p and avoiding $S_{f,g} \setminus \{p\}$.

As the following exercise shows, there is a fundamental relation between residues and tame symbols.

Exercise 7.3 Let $f, g \in \mathbb{C}(X)^\times$ and $p \in X$.

(a) Show that $\mathrm{Res}_p(\eta(f, g)) = 2\pi \log|\partial_p(f, g)|$.
(b) Let z be a local holomorphic coordinate at p, with $z(p) = 0$. For $a, b \in \mathbb{R}$ and $\varepsilon > 0$ sufficiently small, define

$$\gamma_{a,b,\varepsilon}: [a, b] \to X, \quad t \mapsto \varepsilon e^{it}.$$

Show that

$$\lim_{\varepsilon \to 0} \int_{\gamma_{a,b,\varepsilon}} \eta(f,g) = (b-a)\log|\partial_p(f,g)|.$$

(c) Let $\gamma: [0,1] \to X$ be a continuously differentiable path such that $\gamma(0,1) \cap S_{f,g} = \emptyset$. Show that $\int_\gamma \eta(f,g)$ converges absolutely.

(d) Assume that γ has an end point $p \in S_{f,g}$ and that $\eta(f,g)$ has non-zero residue at p. Show that there exist paths γ' homotopic to γ (with the same end points) and satisfying $\gamma'(0,1) \cap S_{f,g} = \emptyset$, such that

$$\int_{\gamma'} \eta(f,g) \neq \int_\gamma \eta(f,g).$$

Hint   (a) Pass to polar coordinates.   □

We finally come to the actual construction of the regulator map. Let $\xi = \sum_j \{f_j, g_j\} \in K_2(\mathbb{C}(X))$ with $f_j, g_j \in \mathbb{C}(X)^\times$. Consider the 1-form

$$\eta = \sum_j \eta(f_j, g_j).$$

It is defined on $Y' = X \setminus S'$, where $S' = \bigcup_j S_{f_j, g_j}$ is finite. Enlarging S' if necessary, we may assume that Y' is contained in Y.

Assume that ξ has trivial tame symbols on Y, in other words, $\partial_p(\xi) = 1$ for every $p \in Y$. By Exercise 7.3(a) we know that η has trivial residue at every point of Y. Although the form η is a priori not regular on Y, we now show that its de Rham cohomology class extends to Y. Indeed, the Mayer–Vietoris sequence [36, Chapter 1, Section 2] applied with small open discs at each point of $S' \setminus S$ provides an exact sequence

$$0 \longrightarrow H^1(Y,\mathbb{R}) \longrightarrow H^1(Y',\mathbb{R}) \xrightarrow{\text{Res}} \bigoplus_{P \in S' \setminus S} \mathbb{R}. \tag{7.6}$$

Since $S' \setminus S$ is contained in Y, we have $\text{Res}(\eta) = 0$, so that η defines a unique element $[\eta] \in H^1(Y,\mathbb{R})$. Importantly, this class $[\eta]$ does not depend on the particular decomposition $\xi = \sum_j \{f_j, g_j\}$, because of the bilinearity of $\eta(f,g)$ and the exactness of $\eta(f, 1-f)$ (Theorem 7.2).

Definition 7.4 Let Y be a smooth complex algebraic curve. The *regulator map* associated to Y, denoted by reg_Y, is the composite map

$$\text{reg}_Y: K_2(Y) \longrightarrow \ker \partial \longrightarrow H^1(Y(\mathbb{C}),\mathbb{R}), \tag{7.7}$$

where the first arrow is given by the localisation sequence (7.5) and the second arrow is the map $\xi \mapsto [\eta]$ defined above.

Exercise 7.4 (a) Derive the exact sequence (7.6) from the Mayer–Vietoris sequence.

(b) What is the image of the residue map Res? Distinguish the cases where Y is compact or not.

The regulator map reg_Y provides a bilinear pairing

$$\langle \cdot, \cdot \rangle \colon H_1(Y(\mathbb{C}), \mathbb{Z}) \times K_2(Y) \to \mathbb{R}, \tag{7.8}$$

where $H_1(Y(\mathbb{C}), \mathbb{Z})$ is the first homology group of $Y(\mathbb{C})$ (see Exercise 7.9). Note that given two invertible functions $f, g \in O(Y)^\times$, we may integrate $\eta(f, g)$ over paths that are not necessarily closed. In Chapter 9, we will consider such integrals where f, g are two modular units on a modular curve, and γ is a modular symbol.

7.3 Relation to the Mahler measure

We will now demonstrate, following Deninger, how the Mahler measure of a two-variable polynomial is related to the regulator map on the associated algebraic curve.

Let $P(x, y) \in \mathbb{C}[x, y]$ be an irreducible polynomial. Let $P^*(x)$ be the leading coefficient of $P(x, y)$ seen as a polynomial in y. For every x on the unit circle $\mathbb{T}^1$ such that $P^*(x) \neq 0$, Jensen's formula with respect to y gives

$$\frac{1}{2\pi i} \int_{|y|=1} \log |P(x, y)| \frac{dy}{y} = \log |P^*(x)| + \sum_{\substack{|y|>1 \\ P(x,y)=0}} \log |y|, \tag{7.9}$$

where the sum is finite and takes into account multiplicities of the roots. Integrating (7.9) over $x \in \mathbb{T}^1$, we get

$$m(P) = m(P^*) + \frac{1}{2\pi i} \int_\gamma \log |y| \frac{dx}{x}, \tag{7.10}$$

where

$$\gamma = \gamma_P = \{(x, y) \in \mathbb{C}^2 : |x| = 1, \ |y| > 1, \ P(x, y) = 0\}$$

is called the *Deninger path* on the algebraic curve $P(x, y) = 0$. It is oriented by means of the canonical orientation of $x \in \mathbb{T}^1$.

Some qualitative comments on the Deninger path are in order. In general, given $x \in \mathbb{T}^1$, there will be several roots of the algebraic equation $P(x, y) = 0$ satisfying $|y| > 1$. So the Deninger path may consist of several components. Furthermore, if we follow these roots as x varies in $\mathbb{T}^1$, then some of them may cross the unit circle. As a consequence, the components will not be closed paths in general. Figure 7.2 illustrates these phenomena by plotting the y-coordinate of the Deninger path for Smyth's polynomial $1 + x + y$ and for the polynomial $1 + x + y + x^2 + xy + y^2$.

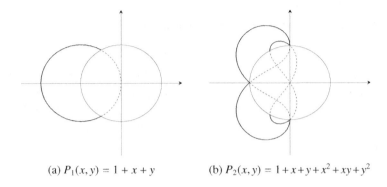

(a) $P_1(x, y) = 1 + x + y$ (b) $P_2(x, y) = 1 + x + y + x^2 + xy + y^2$

Figure 7.2 Illustration of the Deninger path for two polynomials

In order to formulate things more precisely, we introduce some notation. Let $C_P = \{(x, y) \in (\mathbb{C}^\times)^2 : P(x, y) = 0\}$ be the zero locus of P in $(\mathbb{C}^\times)^2$, and let C_P^{reg} denote its smooth part. Note that $\eta(x, y)$ is well defined on the Riemann surface C_P^{reg} and that

$$\eta(x, y)|_{\gamma \cap C_P^{\mathrm{reg}}} = -\log|y|\, d\arg x = i\log|y|\,\frac{dx}{x}.$$

We thus arrive at the following result.

Theorem 7.5 ([68, Proposition 3.3]) *Let $P \in \mathbb{C}[x, y]$ be an irreducible polynomial, and let γ be the associated Deninger path on C_P. Assume that its closure $\bar{\gamma}$ is a finite union of smooth paths contained in C_P^{reg}. Then*

$$\mathrm{m}(P) - \mathrm{m}(P^*) = -\frac{1}{2\pi}\int_{\bar{\gamma}} \eta(x, y).$$

Since the differential form $\eta(x, y)$ is closed, one is free to move the path $\bar{\gamma}$ inside C_P^{reg}, keeping the end points fixed of course.

Furthermore, observe that the boundary of $\bar{\gamma}$ is contained in $\mathbb{T}^2 \cap C_P$. In particular, if P does not vanish on the torus $\mathbb{T}^2$, then γ is the union of *closed* paths, that is, γ defines an element of $H_1(C_P^{\mathrm{reg}}, \mathbb{Z})$.

Example 7.6 It is worth spelling things out for the simple polynomial $P(x, y) = x + y - 1$, whose Mahler measure was evaluated in Section 3.3 (note that $\mathrm{m}(x + y - 1) = \mathrm{m}(x + y + 1)$). The Deninger path $\bar{\gamma}$ consists of those points $(x, 1 - x)$ satisfying $|x| = 1$ and $|1 - x| \geq 1$. This amounts to saying that $x = e^{i\theta}$ with $\theta \in$

$[\frac{\pi}{3}, \frac{5\pi}{3}]$, so that $\overline{\gamma}$ goes from the point $(e^{\pi i/3}, e^{-\pi i/3})$ to $(e^{-\pi i/3}, e^{\pi i/3})$. Moreover, using Theorem 7.2,

$$\eta(x, y) = \eta(x, 1 - x) = d(D(x)).$$

Applying Theorem 7.5, we get

$$m(x + y - 1) = -\frac{1}{2\pi} \int_{\overline{\gamma}} d(D(x)) = -\frac{1}{2\pi}(D(e^{-\pi i/3}) - D(e^{\pi i/3})).$$

Since $D(\overline{z}) = -D(z)$, we get $m(P) = \frac{1}{\pi}D(e^{\pi i/3})$, and we conclude as in Proposition 3.4.

The following exercise shows a variant of Theorem 7.5, using a different Deninger path.

Exercise 7.5 Let $P \in \mathbb{C}[x, y]$ be an irreducible polynomial, and let $P_*(x)$ denote the *lowest*-degree coefficient of $P(x, y)$ with respect to y. Consider

$$\gamma' = \{(x, y) : |x| = 1, \ 0 < |y| < 1, \ P(x, y) = 0\}.$$

Assume that $\overline{\gamma'}$ is a finite union of smooth paths contained in C_P^{reg}. Show that

$$m(P) - m(P_*) = \frac{1}{2\pi} \int_{\gamma'} \eta(x, y).$$

Chapter notes

Historically, the regulator map was introduced in a different way. Given a smooth complex curve Y, one can define a regulator map $\widetilde{\text{reg}}_Y : K_2(Y) \to H^1(Y, \mathbb{C}^\times)$, which lifts the regulator map defined in this chapter in the sense that applying the map $\log |\cdot| : \mathbb{C}^\times \to \mathbb{R}$ recovers reg_Y (see Exercise 7.10). There were definitions of $\widetilde{\text{reg}}_Y$ using various methods by Bloch [27, 25, 82], Deligne (unpublished) and Ramakrishnan [160]. The definition in Exercise 7.10 was given by Beilinson [13].

One may wonder, in Theorem 7.5, how restrictive the assumption that the Deninger path $\overline{\gamma}$ consists of smooth paths avoiding the singular points of C_P is. Using an idea of Bornhorn [31, Lemma 1.7] (see also [98, Lemma 3.1]), one can always find a monomial transformation $(x, y) \mapsto (x^a y^b, x^c y^d)$ with $ad - bc \neq 0$ such that the resulting Deninger path $\gamma' = \{(x, y) : |x^a y^b| = 1, \ |x^c y^d| > 1\}$ avoids the singular points. As we saw in Section 3.5, such a transformation does not change the Mahler measure.

In Section 7.3, the Milnor symbol $\{x, y\}$ defines an element of $K_2(C_P^{\text{reg}})$. One may ask whether this element extends to the smooth compactification $\hat{C}_P$ of

C_p^{reg}. This is important in view of the relation with Beilinson's conjectures, which will be discussed in Section 8.2. We say that $P(x, y)$ is a *tempered polynomial* if $\{x, y\}$ extends to an element of $K_2(\hat{C}_P) \otimes \mathbb{Q}$. It turns out that this condition can be read off very simply on the Newton polygon of $P(x, y)$. Isolating the coefficients of a given side of the Newton polygon produces a univariate polynomial called the side polynomial (see [165, Section 8] for the precise definition). The temperedness condition is then equivalent to asking that all the roots of the side polynomials are roots of unity [165, Section 8].

The differential $\eta(x, y)$ involved in Theorem 7.5 is always closed. When it is exact, we say that $P(x, y)$ is an *exact polynomial*. In this case, the Mahler measure can be expressed in terms of the Bloch–Wigner dilogarithm; see [98] for a general and explicit formula. Exact polynomials are investigated in [98] using two notions from real algebraic geometry: the logarithmic Gauss map and amoebas.

Additional exercises

Exercise 7.6 (a) Show that if F is a finite field then $K_2(F) = 0$.

(b) Construct a surjective homomorphism $K_2(\mathbb{R}) \to \{\pm 1\}$.

(c) Let F be a field equipped with a non-trivial discrete valuation v such that the residue field of v has characteristic $\neq 2$. Show that there exists $x \in F^\times$ such that $\{x, x\} \neq 0$.

Exercise 7.7 (Milnor K-theory) Let F be a field and $n \geq 1$ be an integer. Define the Milnor K-group $K_n^M(F) = (\otimes_{\mathbb{Z}}^n F^\times)/R$, where R is the subgroup generated by the elements $\cdots \otimes x \otimes (1 - x) \otimes \cdots$ with $x \in F \setminus \{0, 1\}$. Denote by $\{x_1, \ldots, x_n\}$ the class of $x_1 \otimes \cdots \otimes x_n$. There is a natural map $K_n^M(F) \to K_n(F)$, where K_n is Quillen's K-group. It is *not* an isomorphism in general for $n \geq 3$.

(a) Show that $\{x_1, \ldots, x_n\}$ is antisymmetric in the variables $x_1, \ldots, x_n$, and that $\{\ldots, x, \ldots, -x, \ldots\} = 0$ for any $x \in F^\times$.

(b) Show that if $x_1 + \cdots + x_n = 1$ then $\{x_1, \ldots, x_n\} = 0$.

(c) Show that if $x_1 + \cdots + x_n = 0$ then $\{x_1, \ldots, x_n\} = 0$.

(d) Let v be a discrete valuation on F. Let $O = \{x \in F : v(x) \geq 0\}$ be the valuation ring, $\mathfrak{m}$ the maximal ideal of O and $k = O/\mathfrak{m}$ the residue field. Show that there exists a unique linear map $\partial_v \colon K_n^M(F) \to K_{n-1}^M(k)$ such that for all $u_1, \ldots, u_{n-1} \in O^\times$ and $x \in F^\times$, we have

$$\partial_v(\{u_1, \ldots, u_{n-1}, x\}) = v(x)\{\bar{u}_1, \ldots, \bar{u}_{n-1}\}.$$

Show that in the case $n = 2$, the map ∂_v is the tame symbol defined in Exercise 7.1.

Hint (d) Choose a uniformiser $\pi \in \mathfrak{m} \setminus \mathfrak{m}^2$. Show that there is a map of graded rings $\theta_\pi : K_*^M(F) \to K_*^M(k)[\varepsilon]/(\varepsilon^2 - \{-1\}\varepsilon)$ such that $\theta_\pi(\{\pi^a u\}) = \{\bar{u}\} + a\varepsilon$ for $a \in \mathbb{Z}$ and $u \in O^\times$. Define ∂_v to be the ε-part of θ_π. □

Exercise 7.8 (a) Compute the residues at $z = 0, 1, \infty$ of the differential form $\eta(z, 1 - z)$ on $\mathbb{P}^1(\mathbb{C}) \setminus \{0, 1, \infty\}$.

(b) Deduce another proof of the fact that the form $\eta(f, 1 - f)$ is exact for any meromorphic function $f \neq 0, 1$ on a compact connected Riemann surface.

The next exercise describes the homology of a compact Riemann surface with punctures.

Exercise 7.9 Recall that the first homology group $H_1(X, \mathbb{Z})$ of a topological space X is the abelian group generated by continuous loops $\gamma : \mathbb{R}/\mathbb{Z} \to X$, modulo the relations given by the homotopy of loops and additivity $[\gamma\gamma'] = [\gamma] + [\gamma']$, where $\gamma\gamma'$ denotes the concatenation of two loops γ and γ' such that $\gamma(0) = \gamma'(0)$.

(a) Let X be a compact Riemann surface and $Y = X \setminus S$ where S is a non-empty finite subset of X. Show that there is an exact sequence of abelian groups

$$0 \longrightarrow \mathbb{Z} \overset{\alpha}{\longrightarrow} \bigoplus_{p \in S} \mathbb{Z} \overset{\beta}{\longrightarrow} H_1(Y, \mathbb{Z}) \longrightarrow H_1(X, \mathbb{Z}) \longrightarrow 0,$$

where α sends 1 to $(1, 1, \ldots, 1)$ and β sends the canonical basis element $\varepsilon_p, p \in S$ to a small loop γ_p around p oriented counterclockwise.

(b) Describe the homology of $\mathbb{C} \setminus \{z_1, \ldots, z_n\}$, where $z_1, \ldots, z_n$ are n distinct points.

(c) Show that if X is a compact Riemann surface with at least one puncture, and $Y = X \setminus S$, we still have an exact sequence

$$0 \longrightarrow \bigoplus_{p \in S} \mathbb{Z} \overset{\beta}{\longrightarrow} H_1(Y, \mathbb{Z}) \longrightarrow H_1(X, \mathbb{Z}) \longrightarrow 0.$$

Exercise 7.10 Let Y be a smooth algebraic curve over $\mathbb{C}$. In this exercise we give Beilinson's definition [13] of the 'enhanced' regulator

$$\widehat{\mathrm{reg}}_Y : K_2(Y) \to H^1(Y, \mathbb{C}^\times).$$

Let $p \in Y$ be a base point. For any $f, g \in O(Y)^\times$ and any continuous path

$\gamma: [0, 1] \to Y$ starting at p, define

$$c(f, g)(\gamma) = \exp\left(\frac{1}{2\pi i}\left(\int_{\gamma} \log f \times d \log g - \log g(p) \times \int_{\gamma} d \log f\right)\right) \in \mathbb{C}^{\times},$$

where $\log f$ and $\log g$ are fixed holomorphic branches of the logarithm continuous along γ, and all the integrals start at p.

(a) Show that if γ is closed then $c(f, g)(\gamma)$ depends only on f and g, and not on the branches of the logarithm chosen.
(b) Show that if γ is the boundary of a 2-simplex in Y, then $c(f, g)(\gamma) = 1$.
(c) Deduce that $c(f, g)$ defines an element of $H^1(Y, \mathbb{C}^{\times})$ (for the definition of the singular cohomology group $H^1(Y, \mathbb{C}^{\times})$, see [99, Section 3.1]).
(d) Let X be the smooth compactification of Y. Show that if γ_q is a small loop around $q \in X \setminus Y$, then $c(f, g)(\gamma_q) \in \mathbb{C}^{\times}$ is the tame symbol of $\{f, g\}$ at q.
(e) Show that $c(f, 1 - f) = 1$ whenever f and $1 - f$ are invertible on Y.
(f) Proceeding as in Section 7.2, construct a map $\widehat{\text{reg}}_Y: K_2(Y) \to H^1(Y, \mathbb{C}^{\times})$.
(g) Show that $\log |\widehat{\text{reg}}_Y| = \text{reg}_Y$.

Hint   (e) Show that $c(f, g)$ is functorial in Y, and reduce to the case $Y = \mathbb{P}^1 \setminus \{0, 1, \infty\}$.   □

8

Deninger's method for multivariate polynomials

In this chapter we outline Deninger's original method for evaluating the Mahler measure in the general setting of multivariate polynomials [68]. We define Deligne–Beilinson cohomology in Section 8.1, discuss Beilinson's conjectures on special values of L-functions in Section 8.2 and give Deninger's method in Section 8.3. In the final section of this chapter we survey known results about links between Mahler measures and L-values.

8.1 Deligne–Beilinson cohomology

The regulator map on K_2 of curves defined in Section 7.2 can be generalised to higher-dimensional varieties. The target of this regulator map is given by *Deligne–Beilinson cohomology*. Very roughly, this cohomology is defined like ordinary de Rham cohomology but using differential forms with *logarithmic singularities at infinity*.

In this section we give a concrete definition of Deligne–Beilinson cohomology with real coefficients. The reader should be warned that this cohomology is classically defined as the hypercohomology of the so-called Deligne–Beilinson complex. The definition given here, due to Burgos [55], has the advantage of being more explicit, as it is the cohomology of a single complex. More details can be found in the appendix of this book, where we explain the construction of a motivic ring spectrum representing Deligne–Beilinson cohomology. This point of view has the advantage of providing a conceptual definition of the regulator map, and also gives immediately the various compatibilities that it enjoys.

Let X be a non-singular complex algebraic variety. By Hironaka's resolution of singularities, X admits a non-singular compactification $\overline{X}$ such that $D = \overline{X} \backslash X$ is a divisor with normal crossings in $\overline{X}$ (see Section A.3 for more details). This means that locally the divisor D is given by a union of coordinate hyperplanes, in other words by an equation of the form $z_1 \cdots z_m = 0$ for suitable analytic local coordinates $z_1, \ldots, z_d$ on $\overline{X}$, where d is the dimension of X and $m \leq d$. Note that a compactification is not unique in general: for example, $X = \mathbb{C}^2$ can be compactified as $\mathbb{P}^2(\mathbb{C})$ or as $\mathbb{P}^1(\mathbb{C}) \times \mathbb{P}^1(\mathbb{C})$. In the first case $D \cong \mathbb{P}^1(\mathbb{C})$, and in the second case D is the union of two copies of $\mathbb{P}^1(\mathbb{C})$ crossing transversally.

Definition 8.1 Let ω be a complex C^∞ differential form on X. We say that ω has logarithmic singularities along D if ω is locally in the algebra generated by the following forms:

- the regular forms on $\overline{X}$,
- $\log|z_i|$, $\frac{dz_i}{z_i}$ and $\frac{d\overline{z_i}}{\overline{z_i}}$ for $1 \leq i \leq m$,

where $z_1 \cdots z_m = 0$ is a local equation for the divisor D.

For example, the differential 2-form $\log|z| \frac{dz}{z} \wedge \frac{d\overline{z}}{\overline{z}}$ on $\mathbb{C}^\times$ has logarithmic singularities along $\{0, \infty\}$ in $\mathbb{P}^1(\mathbb{C})$.

Notation 8.2 For $\Lambda \in \{\mathbb{R}, \mathbb{C}\}$, we denote by $E^n_{\log,\Lambda}(X)$ the space of Λ-valued C^∞ differential n-forms on X with logarithmic singularities along D.

The complex structure on X gives rise to a decomposition

$$E^n_{\log,\mathbb{C}}(X) = \bigoplus_{p+q=n} E^{p,q}_{\log,\mathbb{C}}(X), \tag{8.1}$$

where $E^{p,q}$ denotes the subspace of forms of type (p, q). The exterior derivative $d \colon E^n \to E^{n+1}$ decomposes as $d = \partial + \overline{\partial}$ with $\partial \colon E^{p,q} \to E^{p+1,q}$ and $\overline{\partial} \colon E^{p,q} \to E^{p,q+1}$.

The Deligne–Beilinson cohomology groups can be defined as the cohomology groups of certain complexes $E_j(X)$, which we now define.

Definition 8.3 For any integer $j \geq 0$, we define the complex $E_j(X) = (E_j(X)^n, d^n)_{n \geq 0}$ by

$$E_j(X)^n := \begin{cases} (2\pi i)^{j-1} E^{n-1}_{\log,\mathbb{R}}(X) \cap \left(\bigoplus_{\substack{p+q=n-1 \\ p,q<j}} E^{p,q}_{\log,\mathbb{C}}(X) \right) & \text{if } n \leq 2j-1, \\[2ex] (2\pi i)^j E^n_{\log,\mathbb{R}}(X) \cap \left(\bigoplus_{\substack{p+q=n \\ p,q\geq j}} E^{p,q}_{\log,\mathbb{C}}(X) \right) & \text{if } n \geq 2j, \end{cases}$$

$$d^n\omega := \begin{cases} -\mathrm{pr}_j(d\omega) & \text{if } n < 2j-1, \\ -2\partial\overline{\partial}\omega & \text{if } n = 2j-1, \\ d\omega & \text{if } n \geq 2j, \end{cases}$$

where pr_j denotes the projection $\bigoplus_{p,q} \to \bigoplus_{p,q<j}$.

One may picture the complex $E_j(X)$ as follows:

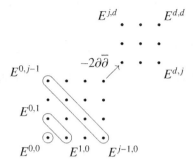

where d is the dimension of X. The term in $E_j(X)$ of degree j is represented here as the diagonal from $E^{j-1,0}$ to $E^{0,j-1}$. The map $-2\partial\bar{\partial}$ is the differential in degree $2j-1$, mapping $E^{j-1,j-1}$ to $E^{j,j}$.

Definition 8.4 Let X be a non-singular complex algebraic variety. The Deligne–Beilinson cohomology groups of X are defined by

$$H^n_{\mathcal{D}}(X, \mathbb{R}(j)) = H^n(E_j(X)) \quad \text{for integers } n, j \geq 0.$$

By results of Burgos [54], the Deligne–Beilinson cohomology groups defined here do not depend on the compactification $\overline{X}$ (see Section A.3 for more details). Moreover, they are finite-dimensional real vector spaces (see the long exact sequence (A.5)).

Here are some examples. For $j = 0$, the complex $E_0(X)$ is simply $E^*_{\log,\mathbb{R}}(X)$. It is known that this complex computes the de Rham cohomology $H^n(X, \mathbb{R})$. The following exercises compute Deligne–Beilinson cohomology in the cases $j > \dim X$ and $j > n$.

Exercise 8.1 Assume $j > \dim X$. Show that the complex $E_j(X)$ is equal to $(2\pi i)^{j-1} E^*_{\log,\mathbb{R}}(X)$ shifted by -1. Deduce the isomorphism $H^n_{\mathcal{D}}(X, \mathbb{R}(j)) = H^{n-1}(X, (2\pi i)^{j-1}\mathbb{R})$.

Exercise 8.2 Show that for $j > n$, we also have

$$H^n_{\mathcal{D}}(X, \mathbb{R}(j)) = H^{n-1}(X, (2\pi i)^{j-1}\mathbb{R}).$$

There is also a version of Deligne–Beilinson cohomology for real varieties, which we now describe. Let X be a non-singular algebraic variety defined over $\mathbb{R}$. The complex points $X(\mathbb{C})$ are endowed with the action of complex conjugation $F_\infty : x \mapsto \overline{x}$. Given a complex differential form ω on $X(\mathbb{C})$, we define its de Rham conjugate by $F_{\mathrm{dR}}(\omega) = F^*_\infty(\overline{\omega})$. The de Rham conjugation is an involution which decides when an *algebraic* differential form on $X_{\mathbb{C}}$ actually comes from $X_{\mathbb{R}}$. The complexes $E_j(X(\mathbb{C}))$ are stable under de Rham conjugation, and we define the complexes $E_j(X_{\mathbb{R}})$ by simply taking the invariants under de Rham conjugation:

$$E_j(X_{\mathbb{R}}) = (E_j(X(\mathbb{C})))^{F_{dR}=1} \quad \text{for } j \geq 0.$$

Definition 8.5 The Deligne–Beilinson cohomology groups of $X_{\mathbb{R}}$ are defined by

$$H^n_{\mathcal{D}}(X_{\mathbb{R}}, \mathbb{R}(j)) = H^n(E_j(X_{\mathbb{R}})) \quad \text{for integers } n, j \geq 0.$$

Note that this is the same as taking the invariants of $H^n_{\mathcal{D}}(X_{\mathbb{C}}, \mathbb{R}(j))$ under de Rham conjugation.

For any non-singular real or complex variety X, there is a well defined cup-product in Deligne–Beilinson cohomology

$$H^n_{\mathcal{D}}(X, \mathbb{R}(j)) \otimes H^{n'}_{\mathcal{D}}(X, \mathbb{R}(j')) \xrightarrow{\cup} H^{n+n'}_{\mathcal{D}}(X, \mathbb{R}(j + j')).$$

The explicit formulae can be found in the appendix (Definition A.22). The cup-product is graded commutative, meaning that $\alpha \cup \alpha' = (-1)^{nn'} \alpha' \cup \alpha$ for any α in $H^n_{\mathcal{D}}(X, \mathbb{R}(j))$ and α' in $H^{n'}_{\mathcal{D}}(X, \mathbb{R}(j'))$. The cup-product is also associative, though not at the level of differential forms if one uses Definition A.22. This is an advantage, as it gives us freedom to choose 'convenient' representatives for the Deligne–Beilinson cohomology classes.

Exercise 8.3 Let X be a non-singular complex algebraic variety.

(a) Let f be an invertible function on X. Show that $\log|f|$ defines a class in $H^1_{\mathcal{D}}(X, \mathbb{R}(1))$.

(b) Let f, g be two invertible functions on X. Using Definition A.22, show that the cohomology class $\log|f| \cup \log|g|$ in $H^2_{\mathcal{D}}(X, \mathbb{R}(2))$ is represented by $i \cdot \eta(f, g)$, where $\eta(f, g)$ is the differential form given in Definition 7.1.

8.2 Beilinson's conjectures

The Beilinson conjectures describe special values of L-functions associated to varieties over number fields (or more generally, to motives) at integer points. The formulation involves the so-called motivic cohomology groups associated to algebraic varieties. Their definition is given in Section A.1.

Let X be a smooth algebraic variety over $\mathbb{R}$ or $\mathbb{C}$. For any $i, j \in \mathbb{Z}$, the motivic cohomology group $H^i_{\mathcal{M}}(X, \mathbb{Q}(j))$ is isomorphic to a certain $\mathbb{Q}$-subspace of Quillen's K-group $K_{2j-i}(X) \otimes \mathbb{Q}$, more precisely the jth eigenspace $K^{(j)}_{2j-i}(X)$ for the Adams operations. Beilinson's regulator, which we define in Sections A.2 and A.3, is a $\mathbb{Q}$-linear map

$$\text{reg}_X^{i,j}: H_{\mathcal{M}}^i(X, \mathbb{Q}(j)) \to H_{\mathcal{D}}^i(X, \mathbb{R}(j)). \tag{8.2}$$

Example 8.6 In the case $i = j = 1$, we have $H_{\mathcal{M}}^1(X, \mathbb{Q}(1)) \cong O(X)^\times \otimes_{\mathbb{Z}} \mathbb{Q}$ (see Section A.1) and the map

$$\text{reg}_X^{1,1}: O(X)^\times \otimes \mathbb{Q} \to H_{\mathcal{D}}^1(X, \mathbb{R}(1))$$

sends any invertible function f to $\log |f|$ (see Exercise A.10).

There is also a cup-product in motivic cohomology. Given invertible functions $f_1, \ldots, f_n \in O(X)^\times \otimes \mathbb{Q}$, we denote their cup-product by

$$\{f_1, \ldots, f_n\} := f_1 \cup \cdots \cup f_n \in H_{\mathcal{M}}^n(X, \mathbb{Q}(n)) \subset K_n(X) \otimes \mathbb{Q}.$$

This generalises (7.4), so that $\{f_1, \ldots, f_n\}$ is also called a Milnor symbol. The regulator map is compatible with taking cup-products, therefore Example 8.6 together with the multiplication formulae in Section A.3 provides us with an explicit representative for the regulator of $\{f_1, \ldots, f_n\}$. For example, if f, g are two invertible functions on a smooth curve Y over $\mathbb{C}$, Exercise 8.3 shows that $\text{reg}_Y^{2,2}\{f, g\}$ is represented by the 1-form $i \cdot \eta(f, g)$. So the regulator map (8.2) is indeed a generalisation of the regulator from Chapter 7.

Let us now assume that X is defined over a number field, and let $X_{\mathbb{R}} = X \times_{\mathbb{Q}} \mathbb{R}$. The regulator map associated to X is defined as the composition

$$\text{reg}_X^{i,j}: H_{\mathcal{M}}^i(X, \mathbb{Q}(j)) \xrightarrow[\text{change}]{\text{base}} H_{\mathcal{M}}^i(X_{\mathbb{R}}, \mathbb{Q}(j)) \xrightarrow{\text{reg}_{X_{\mathbb{R}}}^{i,j}} H_{\mathcal{D}}^i(X_{\mathbb{R}}, \mathbb{R}(j)). \tag{8.3}$$

We now further assume that X is a smooth projective variety defined over $\mathbb{Q}$. We first explain briefly the definition of the L-function associated to the co-homology group $H^i(X)$, where i is any integer satisfying $0 \leq i \leq 2 \dim X$. We refer the reader to [109] for a more complete treatment. Here and in what follows $H^i(X)$ will be purely formal notation, although $H^i(X)$ is in fact an instance of a pure motive [6, 145].

Let p be a prime such that X has good reduction at p, that is, X admits a projective model over $\mathbb{Z}$ whose reduction modulo p is smooth over $\mathbb{F}_p$. Define the Euler factor of $H^i(X)$ at p by

$$P_p(T) = \det(1 - \text{Frob}_p^{-1} \cdot T | H_{\text{ét}}^i(X_{\overline{\mathbb{Q}}}, \mathbb{Q}_\ell)), \tag{8.4}$$

where $\text{Frob}_p \in \text{Gal}(\overline{\mathbb{Q}}/\mathbb{Q})$ denotes a Frobenius element at p, acting on the ℓ-adic étale cohomology group $H_{\text{ét}}^i(X_{\overline{\mathbb{Q}}}, \mathbb{Q}_\ell)$ for a prime $\ell \neq p$. By Deligne's proof of the Weil conjectures [66], the polynomial $P_p(T)$ is in $\mathbb{Z}[T]$ and is independent of $\ell \neq p$. If X has bad reduction at p (there are finitely many such primes p), we define $P_p(T)$ as in (8.4) except that we use the subspace $H^i_{\text{ét}}(X_{\overline{\mathbb{Q}}}, \mathbb{Q}_\ell)^{I_p}$ of invariants under an inertia group I_p in $\text{Gal}(\overline{\mathbb{Q}}/\mathbb{Q})$. It is conjectured that in this case also, $P_p(T)$ is in $\mathbb{Z}[T]$ and is independent of $\ell \neq p$.

The L-function $L(H^i(X), s)$ is then defined, for $s \in \mathbb{C}$ with $\text{Re}(s) \gg 0$, by the Euler product

$$L(H^i(X), s) = \prod_{p \text{ prime}} \frac{1}{P_p(p^{-s})}.$$

(Note that the definition is conjectural because of the bad primes.) By Deligne [66], if we remove the bad Euler factors then $L(H^i(X), s)$ converges for $\text{Re}(s) > i/2 + 1$.

Example 8.7 If $X = \text{Spec}\,\mathbb{Q}$ is a point, then $L(H^0(\text{Spec}\,\mathbb{Q}), s)$ is the Riemann zeta function $\zeta(s)$.

If $X = E$ is an elliptic curve over $\mathbb{Q}$, then $L(H^1(E), s)$ is the Hasse–Weil zeta function $L(E, s)$.

Conjecture 8.8 The function $L(H^i(X), s)$ has a meromorphic continuation to $\mathbb{C}$ and satisfies a functional equation relating the values at s and $i + 1 - s$.

We refer to [67, 5.2] and [109] for the precise statement of the functional equation.

We are now ready to state Beilinson's conjectures. We only deal with values of the L-function in the region of absolute convergence — in this case the formulation is simpler. So let $n > i/2 + 1$ be an integer. The description of the real number $L(H^i(X), n)$ involves the regulator map

$$\text{reg}_X^{i+1,n} : H^{i+1}_{\mathcal{M}}(X, \mathbb{Q}(n)) \to H^{i+1}_{\mathcal{D}}(X_{\mathbb{R}}, \mathbb{R}(n)).$$

Let $H^{i+1}_{\mathcal{M}/\mathbb{Z}}(X, \mathbb{Q}(n))$ be the subspace of $H^{i+1}_{\mathcal{M}}(X, \mathbb{Q}(n))$ defined by Scholl [181], consisting of 'integral' elements (very roughly speaking, those which come from a 'nice' model of X over $\mathbb{Z}$).

Let V be a real vector space. Recall that a $\mathbb{Q}$-structure of V is a $\mathbb{Q}$-subspace S of V spanned by an $\mathbb{R}$-basis of V. This amounts to saying that the canonical map $S \otimes_{\mathbb{Q}} \mathbb{R} \to V$ is an isomorphism of real vector spaces. If V is finite-dimensional and S, S' are two $\mathbb{Q}$-structures of V, then one can define the determinant $\det_S(S')$, which is a non-zero real number well defined up to multiplication by an element of $\mathbb{Q}^\times$. The $\mathbb{Q}$-structures S and S' are called equivalent when $\det_S(S') \in \mathbb{Q}^\times$.

There is a natural $\mathbb{Q}$-structure $\mathcal{R}_{i+1,n}$ in the Deligne–Beilinson cohomology

$H_{\mathcal{D}}^{i+1}(X_{\mathbb{R}}, \mathbb{R}(n))$, coming from the natural $\mathbb{Q}$-structures in singular and de Rham cohomology; see Exercise 8.4(c).

Conjecture 8.9 (Beilinson) Let X be a smooth projective variety over $\mathbb{Q}$, let $0 \le i \le 2 \dim X$ and let $n > i/2 + 1$ be an integer.

(1) The regulator map induces an isomorphism

$$\text{reg}: H_{\mathcal{M}/\mathbb{Z}}^{i+1}(X, \mathbb{Q}(n)) \otimes_{\mathbb{Q}} \mathbb{R} \xrightarrow{\cong} H_{\mathcal{D}}^{i+1}(X_{\mathbb{R}}, \mathbb{R}(n)).$$

(2) We have

$$\det_{\mathcal{R}_{i+1,n}}(\text{reg}\, H_{\mathcal{M}/\mathbb{Z}}^{i+1}(X, \mathbb{Q}(n))) \sim_{\mathbb{Q}^{\times}} L(H^i(X), n),$$

where the notation $a \sim_{\mathbb{Q}^{\times}} b$ means $a/b \in \mathbb{Q}^{\times}$.

Conjecture 8.9(1) predicts, in particular, that motivic cohomology is finite-dimensional. Since this is not known in general and presumably very hard, we often replace the above conjecture by the following weaker one.

Conjecture 8.10 Let $n > i/2 + 1$ be an integer. There exists a $\mathbb{Q}$-subspace W of $H_{\mathcal{M}/\mathbb{Z}}^{i+1}(X, \mathbb{Q}(n))$ such that $\text{reg}(W)$ is a $\mathbb{Q}$-structure of $H_{\mathcal{D}}^{i+1}(X_{\mathbb{R}}, \mathbb{R}(n))$, and we have

$$\det_{\mathcal{R}_{i+1,n}}(\text{reg}(W)) \sim_{\mathbb{Q}^{\times}} L(H^i(X), n).$$

There is another $\mathbb{Q}$-structure $\mathcal{L}_{i+1,n}$ in $H_{\mathcal{D}}^{i+1}(X_{\mathbb{R}}, \mathbb{R}(n))$, where $n > i/2 + 1$ is an integer; see Exercise 8.4(e). In fact, using $\mathcal{L}_{i+1,n}$ instead of $\mathcal{R}_{i+1,n}$ leads to a conjecture for the L-value $L^*(H^i(X), i + 1 - n)$, where the $*$ denotes the leading coefficient of the Taylor expansion at $s = i + 1 - n$. This is the original statement by Beilinson of his conjecture [14, Conjecture 3.4]. The value $L^*(H^i(X), i + 1 - n)$ is on the *left* of the central point $s = (i + 1)/2$, and is related to $L(H^i(X), n)$ by the (conjectural) functional equation. For a detailed discussion of the compatibility of Beilinson's conjecture with the functional equation, see [105, 4.9].

8.3 Deninger's method

Following Deninger's original article [68], we now extend Theorem 7.5 to polynomials in several variables.

Let $P \in \mathbb{C}[x_1, \ldots, x_k]$ be an irreducible polynomial, and let $P^*(x_1, \ldots, x_{k-1})$ be the leading coefficient of P seen as a polynomial in x_k. Using Jensen's formula with respect to x_k, and proceeding as in Section 7.3, we get

$$m(P) - m(P^*) = \frac{1}{(2\pi i)^{k-1}} \int_D \log |x_k| \frac{dx_1}{x_1} \wedge \cdots \wedge \frac{dx_{k-1}}{x_{k-1}}, \qquad (8.5)$$

where

$$D = \{(x_1, \ldots, x_k) : |x_1| = \cdots = |x_{k-1}| = 1, \ |x_k| > 1, \ P(x_1, \ldots, x_k) = 0\} \quad (8.6)$$

is the *Deninger cycle* attached to P. It has real dimension $k - 1$.

Let Z_P^{reg} be the smooth part of the zero locus of P in $(\mathbb{C}^\times)^k$. Note that the differential form on Z_P^{reg},

$$\log |x_k| \, \frac{dx_1}{x_1} \wedge \cdots \wedge \frac{dx_{k-1}}{x_{k-1}},$$

is not necessarily closed. We wish to replace it with a closed differential form, in order to be free to move the cycle D. By reverse induction on $i \in \{1, \ldots, k\}$, the class $\mathrm{reg}_{Z_P^{\text{reg}}}\{x_i, \ldots, x_k\}$ is represented by a form $\eta_i(x_i, \ldots, x_k)$ satisfying

$$\eta_i(x_i, \ldots, x_k)|_D = (-1)^{k-i} \log |x_k| \, \frac{dx_i}{x_i} \wedge \cdots \wedge \frac{dx_{k-1}}{x_{k-1}}.$$

For $i = 1$, we get a differential $(k - 1)$-form $\eta(x_1, \ldots, x_k)$ on Z_P^{reg} representing the regulator of the Milnor symbol $\{x_1, \ldots, x_k\}$. Noting that $k > \dim Z_P^{\text{reg}}$, Exercise 8.1 shows that $H_{\mathcal{D}}^k(Z_P^{\text{reg}}, \mathbb{R}(k))$ coincides with the usual de Rham cohomology, hence η is a closed form. We arrive at the following theorem [68, Proposition 3.3].

Theorem 8.11 *Let $P \in \mathbb{C}[x_1, \ldots, x_k]$ be an irreducible polynomial such that $\overline{D}$ is a topological $(k - 1)$-chain contained in Z_P^{reg}. Then*

$$\mathrm{m}(P) - \mathrm{m}(P^*) = \frac{(-1)^{k-1}}{(2\pi i)^{k-1}} \int_{\overline{D}} \eta(x_1, \ldots, x_k). \quad (8.7)$$

If, moreover, P does not vanish on $\mathbb{T}^k$, then $[D] \in H_{k-1}(Z_P^{\text{reg}}, \mathbb{Z})$ and

$$\mathrm{m}(P) - \mathrm{m}(P^*) = \frac{(-1)^{k-1}}{(2\pi i)^{k-1}} \langle D, \mathrm{reg}_{Z_P^{\text{reg}}}\{x_1, \ldots, x_k\} \rangle. \quad (8.8)$$

When $P \in \mathbb{Q}[x_1, \ldots, x_k]$, Deninger's theorem suggests a relation between $\mathrm{m}(P)$ and some L-value associated to the variety Z_P^{reg}. This would be the L-function in cohomological degree $k - 1$, evaluated at $s = k$. However, there are several important issues:

(1) The variety Z_P^{reg} is not projective, and Beilinson's conjecture is formulated only for smooth projective varieties.
(2) The Deninger cycle $\overline{D}$ may have non-trivial boundary, so it only defines a class in *relative* homology in general.
(3) The right-hand side of (8.8) is only an entry in the matrix representing the Beilinson regulator map, while Beilinson's conjecture is about the determinant of this matrix.

While Theorem 8.11 describes the general situation, an important special case arises when the differential form $\eta(x_1, \ldots, x_k)$ is exact, in which case we say that P is an *exact polynomial*. Writing $\eta(x_1, \ldots, x_k) = d\alpha(x_1, \ldots, x_k)$, Stokes's formula gives

$$m(P) - m(P^*) = \frac{(-1)^{k-1}}{(2\pi i)^{k-1}} \int_{\partial \overline{D}} \alpha(x_1, \ldots, x_k), \qquad (8.9)$$

where $\partial \overline{D}$ is the (oriented) boundary of the Deninger cycle $\overline{D}$. Note that $\partial \overline{D}$ is contained in the torus $\mathbb{T}^k$. Following an idea of Maillot, the boundary $\partial \overline{D}$ is contained in the following algebraic subvariety, called the *Maillot subvariety*:

$$W_P : P(x_1, \ldots, x_k) = \bar{P}\left(\frac{1}{x_1}, \ldots, \frac{1}{x_k}\right) = 0, \qquad (8.10)$$

which suggests, in this case, that $m(P)$ may be related to the cohomology of W_P.

This is indeed so in several examples. Let us look at the simplest case, namely Smyth's polynomial $P(x, y) = 1 + x + y$. Here W_P consists of the two points $(\zeta_3, \bar{\zeta}_3)$ and $(\bar{\zeta}_3, \zeta_3)$, so that the $\mathbb{Q}$-scheme W_P is isomorphic to Spec $\mathbb{Q}(\zeta_3)$. The only interesting cohomology is in degree 0 and the L-function associated to $H^0(W)$ is none other than the Dedekind zeta function of the quadratic field $\mathbb{Q}(\zeta_3)$. And we have $L(H^0(W), 2) = \zeta_{\mathbb{Q}(\zeta_3)}(2) = \zeta(2)L(\chi_{-3}, 2)$. This gives a more conceptual viewpoint on Smyth's formula.

Another example, this time in three variables, is given by the polynomial $P(x, y, z) = (1 + x)(1 + y) + z$. One can prove using Theorem 7.2 that in this case $\eta(x, y, z)$ is exact, and the Maillot subvariety W_P is birational to an elliptic curve E of conductor 15. Boyd has conjectured that the Mahler measure of P is proportional to the L-value of E at $s = 3$; see [118].

Interestingly, Lalín has observed that this exactness property may occur successively [117]: if the differential form α restricted to the subvariety W is exact, then we can apply Stokes again and get an integral on a still smaller subvariety. This is what happens with the polynomial $1 + x + y + z$; see Section 8.4. This process can go on. See also [61, Section 5.2] for a general explanation of Maillot's 'reciprocating' idea.

Question 8.12 The cohomology $H^{k-1}(Z_P^{\text{reg}})$ involved in Deninger's method, and $H^{k-2}(W)$ in the exact case, are all middle-dimensional cohomology groups. Can cohomology in other degrees be related to quantities analogous to the Mahler measure?

In Chapters 9 and 10, we will be concerned with computing integrals like (8.7) when the underlying variety Z_P^{reg} comes (in a precise sense) from modular forms.

8.4 Mahler measures and values of L-functions

In the previous chapters we witnessed numerous connections between Mahler measures and L-functions. In this section, we try to systemise such links by giving a short survey of known results in this direction and outlining how they all fit in the general framework of this chapter.

The first relation between multivariate Mahler measures and values of L-functions was discovered by Smyth (see Proposition 3.4):

$$m(1 + x + y) = \frac{3\sqrt{3}}{4\pi} L(\chi_{-3}, 2) = L'(\chi_{-3}, -1),$$

where χ_{-3} is the non-trivial Dirichlet character of conductor 3. As explained in Example 7.6, the form $\eta(x, y)$ is exact in this case. A few other identities relating Mahler measures and L-values $L(\chi, 2)$, where χ is an odd Dirichlet character, are known; see Exercise 3.4 and [43] for further examples.

Smyth also proved

$$m(1 + x + y + z) = \frac{7}{2\pi^2}\zeta(3) = -14\zeta'(-2)$$

(see Exercise 3.5). Note that in the latter case the Milnor symbol $\{x, y, z\}$ is 2-torsion (see Exercise 7.7), so that the associated form η is again exact. By the work of Lalín [117], one can take $\eta = d\omega$, where ω is exact when restricted to the Maillot subvariety (compare [61, Section 5.2]). This is an example where we have two exactnesses in a row.

In his celebrated paper [39], Boyd made a series of remarkable conjectures for Mahler measures of two-variable polynomials. He discovered several families of (conjectural) identities of the form

$$m(P_k(x, y)) \overset{?}{\sim}_{\mathbb{Q}^\times} L'(E_k, 0),$$

where $P_k(x, y)$ is an integer polynomial in two variables depending on an integer parameter k, and E_k is the elliptic curve defined by $P_k(x, y) = 0$. These identities are in line with Deninger's theorem and Beilinson's conjectures: at least in some cases, the Boyd conjectures can be shown to be a consequence of the Beilinson conjectures.

We have already outlined a portion of Boyd's paper, in particular many families from [39], from a hypergeometric point of view in Section 5.2. Turning back to the famous example $\mu(k) = \mu_2(k) = m(x + 1/x + y + 1/y + k)$, Boyd's conjecture predicts $\mu(k)/L'(E_k, 0) \in \mathbb{Q}^\times$ for every $k \in \mathbb{Z}$, $k \neq 0, \pm 4$, where E_k is the elliptic curve with affine equation $x + 1/x + y + 1/y + k = 0$. Rodriguez Villegas proved that for every $k \in \mathbb{C}$, the measure $\mu(k)$ can be expressed as an Eisenstein–Kronecker series of level $\Gamma_1(4)$. Using a result by Bloch, this allowed him to prove Boyd's conjecture in cases where E_k has complex multiplication (CM), and also in some non-CM case (with an undetermined rational factor) [164]. To our knowledge, the conjecture about $\mu(k)$ in the non-CM case for integer k has been proved only when $k \in \{1, 2, 3, 5, 8, 12, 16\}$, using either an explicit version of Beilinson's theorem [48, 137] or the Rogers–Zudilin method [169, 170, 223, 51]. We will explain the latter method in Chapter 9. Another family studied by Boyd is $n(k) := m(y^2 + kxy + y - x^3)$. This is the general equation (called the Deuring form) for elliptic curves having a rational torsion point of order 3. For this family, the relevant conjecture is known only for $k \in \{1, 2, 3\}$. It is a major challenge to prove Boyd's conjecture for infinitely many values of the parameter k.

Boyd [39], Bertin and Zudilin [22, 23] also investigated families of curves of genus 2. Here the Mahler measure sometimes appears to be proportional to the L-value of an elliptic curve which is obtained as a factor of the Jacobian of the curve.

Regarding polynomials in at least three variables, a few identities are known — see Section 5.3. Bertin [20] managed to generalise Rodriguez Villegas's method to families of polynomials $Q_k(x, y, z)$ defining $K3$ surfaces, expressing $m(Q_k)$ as Eisenstein–Kronecker series. This enabled her to relate $m(Q_k)$ to the L-value of the surface at $s = 3$ for some values of k. Samart also proved results on Mahler measures of families of $K3$ surfaces [172, 173], and with Papanikolas and Rogers even did the case of a Calabi–Yau threefold [150]. Recently, Brunault and Neururer [53] gave a 'modular proof' of Bertin's theorem on $m(Q_2)$, building on the Rogers–Zudilin method. This uses the interpretation of the surface $Q_2(x, y, z) = 0$ as the universal elliptic curve for the group $\Gamma_1(8)$. In this case, the L-function of the surface coincides with the L-function of a modular form of weight 3.

Lalín also developed a new method to express Mahler measures of some families of polynomials in terms of polylogarithms [117], yielding for example

$$m((1 + z_1)(1 + z_2)(1 + z_3) + (1 - z_1)(1 - z_2)(z_4 + z_5)) = \frac{93}{\pi^4}\zeta(5) = 124\zeta'(-4).$$

In Chapter 6 we discussed the family $m(x_1 + x_2 + \cdots + x_k) = m(1 + x_1 + \cdots + x_{k-1})$ of 'linear' Mahler measures originating in the work of Rodriguez Villegas. His investigations led to the remarkable numerical relations, in the cases $k = 5$ and $k = 6$, between the Mahler measures and L-values of certain cusp forms of weight $k - 2$; the details are described in Section 6.2.

Another fascinating conjecture by Boyd (building on Maillot's idea) already discussed in Section 6.3 is

$$m((1 + x)(1 + y) + z) \stackrel{?}{=} -2L'(E, -1), \tag{8.11}$$

where E is an elliptic curve of conductor 15. The surprising appearance of E is explained using the framework of Section 8.3, once we know that the Maillot subvariety

$$\begin{cases} (1+x)(1+y) + z = 0, \\ (1+x^{-1})(1+y^{-1}) + z^{-1} = 0 \end{cases}$$

is birational to E. Conditionally on the conjecture that $K_4(E)$ has rank 1 and up to an undetermined rational factor, the identity (8.11) was proved by Lalín [118].

Chapter notes

We explained in Section 8.3 how exact polynomials in two variables can be generalised in the multivariate setting. Doran and Kerr [77] generalised the notion of temperedness for polynomials with an arbitrary number of variables, using Newton polytopes.

There is a version of Deligne–Beilinson cohomology with integral coefficients. Contrary to what one may expect, it is usually bigger than the cohomology with $\mathbb{R}$-coefficients. For example, if Y is a smooth curve then $H^2_{\mathcal{D}}(Y, \mathbb{Z}(2))$ is isomorphic to $H^1(Y, \mathbb{C}^\times)$, which contains $H^1(Y, \mathbb{R})$ as a direct summand. Bloch constructed a regulator map from higher Chow groups to integral Deligne cohomology using Chern classes [26]. In the case of curves, it agrees with the enhanced regulator $K_2(Y) \to H^1(Y, \mathbb{C}^\times)$ from Exercise 7.10. In view of these refinements, one may wonder whether there is a sufficiently natural 'enhanced' Mahler measure $\hat{M}(P) \in \mathbb{C}^\times$ (maybe depending on some choices) such that $|\hat{M}(P)| = M(P)$, and thus $\log |\hat{M}(P)| = m(P)$.

By work of Suslin and Voevodsky [213, 136], motivic cohomology (of schemes of finite type over a field) can also be defined with integral coefficients. Cisinski and Déglise have extended this to general schemes [59]. Voevodsky proved that for smooth varieties, his motivic cohomology groups coincide with Bloch's higher Chow groups [212].

Good surveys on Beilinson's conjectures and known results about it include (in chronological order) [179, 161, 70, 146]. Beilinson has proved that Conjecture 8.10 holds for L-functions of modular forms of weight 2 at any integer $n \geq 2$. This was later extended by Deninger and Scholl (unpublished) and Gealy [91] to L-values of modular forms of arbitrary weight $k \geq 2$ at any integer $n \geq k$. (We will formally introduce modular forms and their L-functions in Chapter 9.)

Beilinson's conjectures can be generalised in two directions. First, they can be extended to pure motives $H^i(X)$ endowed with an action of a semisimple $\mathbb{Q}$-algebra (see [105]). This is in essence already in Beilinson's article [15], where he considers modular curves endowed with the algebra of Hecke correspondences. Second, the Bloch–Kato conjecture pins down (up to sign) the rational factor appearing in Beilinson's conjecture [28, 87]. The common generalisation of these deep ideas is the equivariant Tamagawa number conjecture (ETNC) by Burns and Flach [86].

Additional exercises

The following exercise strengthens Exercise 8.2 for smooth projective varieties, and gives the definition of the $\mathbb{Q}$-structures appearing in the Beilinson conjecture.

Exercise 8.4 Let X be a non-singular projective complex variety. As shown in the appendix (A.5), for any $j \geq 0$ there is a long exact sequence

$$\cdots \to H^{n-1}(X, \mathbb{C})/F^j H^{n-1}(X, \mathbb{C}) \to H^n_{\mathcal{D}}(X, \mathbb{R}(j)) \to H^n(X, \mathbb{R}(j))$$
$$\to H^n(X, \mathbb{C})/F^j H^n(X, \mathbb{C}) \to \cdots,$$

where $H^n(X, \cdot)$ denotes de Rham cohomology, $\mathbb{R}(j) = (2\pi i)^j \mathbb{R}$ denotes the Tate twist and F is the Hodge filtration, given by

$$F^k H^n(X, \mathbb{C}) = \bigoplus_{\substack{p+q=n \\ p \geq k}} H^{p,q}(X).$$

We assume throughout the exercise that $n < 2j$.

(a) Show that the map $H^n(X, \mathbb{R}(j)) \to H^n(X, \mathbb{C})/F^j H^n(X, \mathbb{C})$ is injective.

(b) Deduce that for $n < 2j$, we have a short exact sequence

$$0 \to H^{n-1}(X, \mathbb{R}(j)) \to H^{n-1}(X, \mathbb{C})/F^j H^{n-1}(X, \mathbb{C}) \to H^n_{\mathcal{D}}(X, \mathbb{R}(j)) \to 0.$$
$$(8.12)$$

We recall Grothendieck's theorem [96]: the algebraic de Rham cohomology $H_{\mathrm{dR}}(X)$ is isomorphic to the de Rham cohomology $H^{\cdot}(X, \mathbb{C})$.

(c) In the case that X is defined over $\mathbb{Q}$, use (8.12) to define a natural $\mathbb{Q}$-structure (up to equivalence) $\mathcal{R}_{n,j}$ in the real vector space $H^n_{\mathcal{D}}(X_{\mathbb{R}}, \mathbb{R}(j))$.

(d) Construct another short exact sequence

$$0 \to F^j H^{n-1}(X, \mathbb{C}) \to H^{n-1}(X, \mathbb{R}(j-1)) \to H^n_{\mathcal{D}}(X, \mathbb{R}(j)) \to 0. \quad (8.13)$$

(e) In the case that X is defined over $\mathbb{Q}$, deduce another $\mathbb{Q}$-structure (up to equivalence) $\mathcal{L}_{n,j}$ in $H_{\mathcal{D}}^n(X_{\mathbb{R}}, \mathbb{R}(j))$.

(f) Compare the $\mathbb{Q}$-structures $\mathcal{R}_{n,j}$ and $\mathcal{L}_{n,j}$ in the case $X = \operatorname{Spec}\mathbb{Q}$, $n = 1$ and $j \geq 1$ arbitrary. Check that this is compatible with the functional equation of the Riemann zeta function [109, I.1.2], as explained at the end of Section 8.2.

(g) Compare the $\mathbb{Q}$-structures $\mathcal{R}_{n,j}$ and $\mathcal{L}_{n,j}$ in the case that $X = E$ is an elliptic curve defined over $\mathbb{Q}$, $n = 2$ and $j \geq 2$ arbitrary. Show that this is compatible with the functional equation of $L(E, s)$, which we recall here: defining $\Lambda(E, s) = N^{s/2}(2\pi)^{-s}\Gamma(s)L(E, s)$ with $N \geq 1$ being the conductor of E, we have $\Lambda(E, s) = w(E)\Lambda(E, 2 - s)$ where $w(E) = \pm 1$ is the root number of E.

Hint (g) Writing E in short Weierstrass form $y^2 = x^3 + ax + b$, the algebraic de Rham cohomology $H_{dR}^1(E)$ is spanned by dx/y and $x\,dx/y$, whose integrals along a basis of $H_1(E, \mathbb{Z})$ are given respectively by the periods and quasi-periods of E; see [192, I.5]. $\qquad\square$

9

The Rogers–Zudilin method

Our principal goal in this chapter is the Mahler measure of the Boyd–Deninger polynomial $P(x, y) = x + 1/x + y + 1/y + 1$.

Theorem 9.1 (Rogers–Zudilin [170]) *We have*

$$m(P) = L'(E, 0) = \frac{15}{4\pi^2} L(E, 2),$$

where E is the elliptic curve defined as the projective closure of the curve $P(x, y) = 0$.

This is actually one of the first non-CM cases of Boyd's conjectures solved by Rogers and Zudilin.

After a crash course on modular curves and modular forms, we apply the Deninger method (from Section 7.3) to the polynomial $P(x, y)$ defining the modular curve $X_1(15)$. Furthermore, we explain the Rogers–Zudilin method and execute it on the integral obtained from Theorem 7.5 to express the Mahler measure $m(P)$ through the L-value of $X_1(15)$ predicted by Beilinson's conjecture.

One warning: the proof of Theorem 9.1 given in [170] is very different from the one in this chapter. Though also based on the technique we explain in Section 9.3 below, it uses hypergeometric expressions of $m(P)$ discussed in Chapter 5 and modular equations instead of Deninger's argument.

9.1 Modular curves, modular units and modular forms

In this section we give a very brief introduction to modular curves and modular forms.

Let $\mathcal{H} = \{\tau \in \mathbb{C} : \text{Im}\, \tau > 0\}$ be the Poincaré upper half-plane. The group

$SL_2(\mathbb{R})$ acts on $\mathcal{H}$ by Möbius transformations, $\tau \mapsto (a\tau + b)/(c\tau + d)$ for any $\left(\begin{smallmatrix} a & b \\ c & d \end{smallmatrix}\right)$ in $SL_2(\mathbb{R})$.

Definition 9.2 Let $N \geq 1$ be an integer. The congruence groups $\Gamma(N)$, $\Gamma_1(N)$ and $\Gamma_0(N)$ are defined by

$$\Gamma(N) = \{g \in SL_2(\mathbb{Z}) : g \equiv \left(\begin{smallmatrix} 1 & 0 \\ 0 & 1 \end{smallmatrix}\right) \bmod N\},$$

$$\Gamma_1(N) = \{g \in SL_2(\mathbb{Z}) : g \equiv \left(\begin{smallmatrix} 1 & * \\ 0 & 1 \end{smallmatrix}\right) \bmod N\},$$

$$\Gamma_0(N) = \{g \in SL_2(\mathbb{Z}) : g \equiv \left(\begin{smallmatrix} * & * \\ 0 & * \end{smallmatrix}\right) \bmod N\}.$$

The associated modular curves $Y_*(N)$ for $* \in \{\emptyset, 1, 0\}$ are defined as the quotient Riemann surfaces $\Gamma_*(N) \backslash \mathcal{H}$.

The quotient $\Gamma_*(N)\backslash\mathcal{H}$ makes sense as a Riemann surface because the action of $SL_2(\mathbb{Z})$ on $\mathcal{H}$ is holomorphic and proper. This last point follows from the isomorphism $\mathcal{H} \cong SL_2(\mathbb{R})/SO(2)$, the discreteness of $SL_2(\mathbb{Z})$ in $SL_2(\mathbb{R})$ and the compactness of the special orthogonal group $SO(2)$.

It turns out that these Riemann surfaces can be compactified by adding the finite set of cusps $\Gamma_*(N) \backslash \mathbb{P}^1(\mathbb{Q})$, where $\mathbb{P}^1(\mathbb{Q}) = \mathbb{Q} \cup \{\infty\}$. The compactification of $Y_*(N)$ is denoted by $X_*(N)$. This implies that the Riemann surfaces $Y_*(N)$ and $X_*(N)$ are complex algebraic curves. They can actually be defined over $\mathbb{Q}$ — a fact of great importance in number theory.

For example, let us describe the cusp ∞ of the modular curve $X_1(N)$ (in what follows we deal with the group $\Gamma_1(N)$, but the discussion extends to the other groups). Since $\Gamma_1(N)$ contains the matrix $\left(\begin{smallmatrix} 1 & 1 \\ 0 & 1 \end{smallmatrix}\right)$, every $\Gamma_1(N)$-invariant function on $\mathcal{H}$ is invariant by $\tau \to \tau + 1$. So every holomorphic function on $Y_1(N)$ is actually a holomorphic function of $q = e^{2\pi i \tau}$, with $0 < |q| < 1$. Note that q tends to 0 when τ tends to ∞ on the positive imaginary axis. So we take q as a local holomorphic coordinate at ∞. A complete system of neighbourhoods of the cusp ∞ is given by $\{\text{Im}(\tau) \geq y_0\}$ with $y_0 > 0$. In terms of the q-coordinate, this corresponds to the usual system of neighbourhoods at $q = 0$.

Given two cusps $\alpha, \beta \in \mathbb{P}^1(\mathbb{Q})$, the *modular symbol* $\{\alpha, \beta\}$ is the hyperbolic geodesic from α to β in $\mathcal{H}$ (see Figure 9.1). We also denote by $\{\alpha, \beta\}$ its image in $X_1(N)$. It is an element of the relative homology group $H_1(X_1(N), \{\text{cusps}\}, \mathbb{Z})$.

Definition 9.3 A *modular unit* on $Y_1(N)$ is a meromorphic function on $X_1(N)$ whose only zeros and poles are at the cusps. In other words, it is an element of $O(Y_1(N))^\times$.

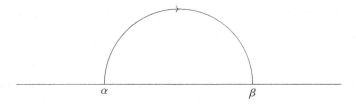

Figure 9.1 Modular symbol

Examples of modular units are given, for any $a \in \{1, \ldots, N-1\}$, by

$$\tilde{g}_a(\tau) = q^{NB_2(a/N)/2} \prod_{\substack{n \geq 1 \\ n \equiv a \bmod N}} (1-q^n) \prod_{\substack{n \geq 1 \\ n \equiv -a \bmod N}} (1-q^n), \quad \text{where } \tau \in \mathcal{H}, \ q = e^{2\pi i \tau}.$$

(9.1)

Here $B_2(x) = x^2 - x + \frac{1}{6}$ is the Bernoulli polynomial and the fractional power q^λ is defined by $q^\lambda = \exp(2\pi i \lambda \tau)$ for any $\lambda \in \mathbb{Q}$. The products are taken over the positive integers n congruent to a (respectively, $-a$) mod N. Note that $|q| < 1$, so that these infinite products are well defined and converge rapidly. By definition $\tilde{g}_a$ is non-vanishing on $\mathcal{H}$, and one can show that $\tilde{g}_a$ defines an element of $O(Y_1(N))^\times \otimes_{\mathbb{Z}} \mathbb{Q}$ (Exercise 9.9). Concretely, this means that some power of $\tilde{g}_a$ is a modular unit on $Y_1(N)$. These units are particular cases of the so-called Siegel units $g_{a,b}$; see Exercise 9.8.

We give a brief definition of modular forms. For any integer $k \in \mathbb{Z}$, we define a right action of $\mathrm{SL}_2(\mathbb{R})$ on the space of holomorphic functions $f \colon \mathcal{H} \to \mathbb{C}$ by

$$(f|_k\gamma)(\tau) = (c\tau + d)^{-k} f(\gamma\tau), \quad \text{where } \gamma = \begin{pmatrix} a & b \\ c & d \end{pmatrix} \in \mathrm{SL}_2(\mathbb{R}).$$

Definition 9.4 A *modular form* of weight $k \geq 1$ on $\Gamma_1(N)$ is a holomorphic function $f \colon \mathcal{H} \to \mathbb{C}$ such that

- for every $\gamma \in \Gamma_1(N)$, we have $f|_k\gamma = f$,
- for every $\gamma \in \mathrm{SL}_2(\mathbb{Z})$, the function $(f|_k\gamma)(\tau)$ is bounded when $\mathrm{Im}(\tau) \to +\infty$.

If in addition $(f|_k\gamma)(\tau) \to 0$ when $\mathrm{Im}(\tau) \to +\infty$ for every $\gamma \in \mathrm{SL}_2(\mathbb{Z})$, we say that f is a *cusp form*.

We denote by $M_k(\Gamma_1(N))$ the space of modular forms of weight k on $\Gamma_1(N)$, and by $S_k(\Gamma_1(N))$ the subspace of cusp forms.

Since $\begin{pmatrix} 1 & 1 \\ 0 & 1 \end{pmatrix} \in \Gamma_1(N)$, every $f \in M_k(\Gamma_1(N))$ is invariant under $\tau \mapsto \tau + 1$, and hence admits a *Fourier expansion* $f(\tau) = \sum_{n=0}^\infty a_n e^{2\pi i n \tau}$. In the case of modular forms on $\Gamma(N)$, we get a Fourier expansion of the form $f(\tau) = \sum_{n=0}^\infty a_n e^{2\pi i n \tau / N}$

instead. We abbreviate these expansions by writing $f = \sum a_n q^n$ (respectively, $f = \sum a_n q^{n/N}$).

The space $M_k(\Gamma_1(N))$ is a finite-dimensional complex vector space. As a consequence, proving that two modular forms are equal can be done by checking that sufficiently many Fourier coefficients coincide. In fact, it is enough to check the Fourier coefficients a_n with $0 \leq n \leq \lfloor km/12 \rfloor$, where m is the index of $\Gamma_1(N)$ in $SL_2(\mathbb{Z})$; this is known as the Sturm bound [204] (see also [201]).

A *congruence subgroup* is a subgroup of $SL_2(\mathbb{Z})$ containing $\Gamma(N)$ for some $N \geq 1$. In a completely similar way, one defines modular curves, modular units and modular forms for arbitrary congruence subgroups.

Exercise 9.1 Show that the complex vector space $S_2(\Gamma_1(N))$ is isomorphic to the space $\Omega^1(X_1(N))$ of holomorphic differential 1-forms on the compact Riemann surface $X_1(N)$.

Hint Send f to $f(\tau)\,d\tau$.   □

Exercise 9.2 Show that for any modular unit $u \in O(Y_1(N))^\times$, we have

$$\frac{d}{d\tau} \log u(\tau) \in M_2(\Gamma_1(N)).$$

An important property of modular forms is that their Fourier coefficients grow at most polynomially. More precisely, let f be a modular form of weight $k \geq 1$ on some congruence subgroup. If f is a cusp form, Hecke showed that $|a_n(f)| \ll n^{k/2}$ (see [142, Corollary 2.1.6]). In the case that f is an Eisenstein series (see Section 10.4 for the definitions), a direct inspection of the Fourier coefficients gives $|a_n(f)| \ll n^{k-1}$ if $k \geq 3$, and $|a_n(f)| \ll n^{k-1+\varepsilon}$ for any $\varepsilon > 0$ if $k = 1, 2$ (see [142, Theorem 4.7.3]). The space of modular forms decomposes as the direct sum of the subspaces of cusp forms and Eisenstein series, which proves the claim on the coefficient growth.

Definition 9.5 Let $f(\tau) = \sum_{n=0}^{\infty} a_n q^{n/N}$ be a modular form in $M_k(\Gamma(N))$ with $k \geq 1$ and $N \geq 1$. The *L-function* associated to f is the Dirichlet series

$$L(f, s) = \sum_{n=1}^{\infty} \frac{a_n}{(n/N)^s}, \quad \text{where } s \in \mathbb{C},\ \mathrm{Re}\, s \gg 0.$$

The *completed L-function of f* is defined by

$$\Lambda(f, s) = N^{s/2}(2\pi)^{-s}\Gamma(s)L(f, s), \quad \text{where } s \in \mathbb{C},\ \mathrm{Re}\, s \gg 0.$$

Since the Fourier coefficients grow at most polynomially, the Dirichlet series $L(f, s)$ converges and is holomorphic when $\mathrm{Re}(s)$ is sufficiently large. The theory of Mellin transforms enables one to write the completed L-function as an integral

$$\Lambda(f, s) = N^{s/2} \int_0^\infty (f(iy) - a_0) y^s \frac{dy}{y}.$$

From there, we can show that $\Lambda(f, s)$ has a meromorphic continuation to $\mathbb{C}$, with at most simple poles at $s = 0$ and $s = k$ (see Exercise 9.13). The residue of $\Lambda(f, s)$ at $s = 0$ is equal to $-a_0$, so we define the regularised value

$$\Lambda^*(f, 0) = \lim_{s \to 0}\left(\Lambda(f, s) + \frac{a_0}{s}\right).$$

9.2 Deninger's method applied to $X_1(15)$

Let $P(x, y) = x + 1/x + y + 1/y + 1$. We first determine the Deninger path $\gamma = \{(x, y) : |x| = 1, |y| > 1, P(x, y) = 0\}$.

Exercise 9.3 Let Z_P be the zero locus of P in $(\mathbb{C}^\times)^2$.

(a) Show that Z_P is a smooth algebraic curve.
(b) Show that the Deninger path γ is given by

$$\gamma = \left\{(e^{i\theta}, Y(\theta)) : \theta \in \left(-\frac{\pi}{3}, \frac{\pi}{3}\right), Y(\theta) = -\cos\theta - \frac{1}{2} - \sqrt{\left(\cos\theta + \frac{1}{2}\right)^2 - 1}\right\}.$$

Next we determine the compactification of Z_P.

Exercise 9.4 (a) Show that the projective closure $\overline{Z_P}$ of Z_P in $\mathbb{P}^2(\mathbb{C})$ is a non-singular cubic defined over $\mathbb{Q}$.

(b) Show that $E = \overline{Z_P}$ has a rational point, hence is an elliptic curve defined over $\mathbb{Q}$.

(c) Show that $E \setminus Z_P$ is a subgroup of $E(\mathbb{Q})$ isomorphic to $\mathbb{Z}/4\mathbb{Z}$.

The elliptic curve E has conductor 15 and, by the modularity theorem, we know that there exists a non-constant morphism $X_1(15) \to E$. In fact, we do not need the modularity theorem and can construct directly a modular parametrisation of E as follows.

Lemma 9.6 *Let u, v be the modular units on $Y_1(15)$ defined by $u = \tilde{g}_8/\tilde{g}_2$ and $v = -\tilde{g}_{11}/\tilde{g}_1$. Then we have $u + 1/u + v + 1/v + 1 = 0$.*

Sketch of proof Using Exercise 9.9 we get $u, v \in O(Y_1(15))^\times$. The function $h = u + 1/u + v + 1/v + 1$ is holomorphic on $Y_1(15)$ and meromorphic at the cusps. Using a computer and Exercises 9.8, 9.9, we bound the order of the possible pole of h at each cusp different from ∞. Finally, we check that h

vanishes at ∞ with sufficiently high order. Since h has as many zeros as poles on $X_1(15)$, this ascertains that $h = 0$.   □

Now we may consider the map

$$\varphi_0 : Y_1(15) \to Z_P, \quad \tau \mapsto (u(\tau), v(\tau)).$$

This map extends to a holomorphic map $\varphi : X_1(15) \to E$, which is our modular parametrisation. One can show that $X_1(15)$ has genus 1, hence is an elliptic curve.

Exercise 9.5 Determine $\varphi(\infty)$ and show that φ is unramified at ∞.

Corollary 9.7 *The map φ is an isomorphism.*

It turns out that the end points of the Deninger path $\bar{\gamma}$ are cusps on $X_1(15)$. More precisely, using Exercise 9.8 one can show that $\varphi(1/5) = (e^{-\pi i/3}, -1)$ and $\varphi(-1/5) = (e^{\pi i/3}, -1)$. With some more effort, one can show that $\bar{\gamma}$ is in fact homologous to $\varphi(\{1/5, -1/5\})$. This is not obvious: given two points p and q on the elliptic curve E, there are many paths going from p to q up to homotopy. We solve the problem by considering an invariant differential form ω on E and computing numerically the integrals of ω along the modular symbol and the Deninger path. The difference between these two integrals belongs to the period lattice of ω, and hence can be ascertained to be zero by a finite computation. It would be interesting, nevertheless, to devise a method not relying on numerical approximation.

Using Deninger's theorem, we get

$$\mathrm{m}(P) = -\frac{1}{2\pi} \int_{\bar{\gamma}} \eta(x, y) = -\frac{1}{2\pi} \int_{1/5}^{-1/5} \eta(u, v). \tag{9.2}$$

By Exercise 7.3(c), the integral $\int_\alpha^\beta \eta(u_1, u_2)$ converges absolutely for any modular units u_1, u_2 and any $\alpha, \beta \in \mathbb{P}^1(\mathbb{Q})$, so that the right-hand side of (9.2), as well as the subsequent integrals, makes sense.

We now want to decompose the modular symbol $\{1/5, -1/5\}$ as $\{1/5, \infty\} + \{\infty, -1/5\}$; see the illustration in Figure 9.2. However, one should be careful that for general modular units,

$$\int_\alpha^\gamma \eta(u, v) \neq \int_\alpha^\beta \eta(u, v) + \int_\beta^\gamma \eta(u, v)!$$

The reason is that the three hyperbolic geodesics involved approach the three cusps from different directions, and $\eta(u, v)$ may have non-trivial residues at these cusps. In our case, however, $\eta(u, v)$ is regular at $\pm 1/5$, and a simple computation shows that $\partial_\infty \eta(u, v) = 1$ (see Exercise 9.10(a)), hence the residue of

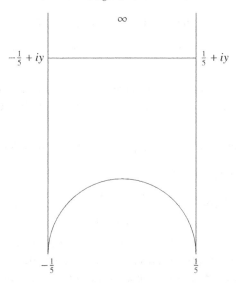

Figure 9.2 Integration path

$\eta(u, v)$ at infinity is 0. By Exercise 7.3, the limit of $\int_{-1/5+iy}^{1/5+iy} \eta(u, v)$ as $y \to +\infty$ is proportional to this residue, hence vanishes. Since $\eta(u, v)$ is closed, using Stokes' formula for the truncated domain given in Figure 9.2 and taking the limit gives

$$\mathrm{m}(P) = -\frac{1}{2\pi}\left(\int_{1/5}^{\infty} - \int_{-1/5}^{\infty} \right) \eta(u, v). \tag{9.3}$$

9.3 The Rogers–Zudilin trick

We will compute, more generally, for any $N \geq 1$, any $a, b \in \{1, \ldots, N-1\}$ and $c \in \mathbb{Z}$, the integral of $\eta(\tilde{g}_a, \tilde{g}_b)$ over $\{c/N, \infty\}$. Taking the logarithm of $\tilde{g}_a$ gives

$$\log \tilde{g}_a(\tau) = \pi i N B_2\left(\frac{a}{N}\right)\tau + \sum_{\substack{n \geq 1 \\ n \equiv a \bmod N}} \log(1 - q^n) + \sum_{\substack{n \geq 1 \\ n \equiv -a \bmod N}} \log(1 - q^n)$$

$$= \pi i N B_2\left(\frac{a}{N}\right)\tau - \sum_{\substack{m, n \geq 1 \\ n \equiv a \bmod N}} \frac{q^{mn}}{m} - \sum_{\substack{m, n \geq 1 \\ n \equiv -a \bmod N}} \frac{q^{mn}}{m}.$$

Taking the real and imaginary parts, we may compute $\log|\tilde{g}_a|$ as well as the logarithmic derivative $\mathrm{d} \arg \tilde{g}_a = \mathrm{Im}(\mathrm{d} \log \tilde{g}_a)$.

Exercise 9.6 Show that there exist functions $\alpha, \beta \colon \mathbb{Z}/N\mathbb{Z} \to \mathbb{R}$ such that

$$\mathrm{d}\arg \tilde{g}_a(c/N + iy) = \sum_{m,n \geq 1} \alpha(m)\beta(n)ne^{-2\pi mny}\mathrm{d}y.$$

Regarding $\log |\tilde{g}_a|$, we will need an alternative expression involving $1/y$ instead of y in the exponents. Using the Siegel units from Exercise 9.8, we may write

$$|\tilde{g}_a(c/N + iy)| = |g_{a,ac}(iNy)| = \left| g_{ac,-a}(i/(Ny)) \right|.$$

Taking the logarithm, we get

$$\log |\tilde{g}_a(c/N + iy)| = \frac{c_1}{y} + \sum_{m,n \geq 1} \alpha'(m)\beta'(n)m^{-1}e^{-\frac{2\pi mn}{N^2 y}}$$

for some constant $c_1 \in \mathbb{R}$ and functions $\alpha', \beta' \colon \mathbb{Z}/N\mathbb{Z} \to \mathbb{R}$.

Now let us integrate $\eta(\tilde{g}_a, \tilde{g}_b)$ over $\{c/N, \infty\}$. We will deal only with the term $\eta_1 = \log |\tilde{g}_a| \times \mathrm{d}\arg(\tilde{g}_b)$, as the other term can be treated similarly. Also, we will ignore the constant term c_1/y in $\log |\tilde{g}_a|$. Then we can write

$$\eta_1(c/N + iy) = \sum_{m_1,n_1,m_2,n_2 \geq 1} \alpha_1(m_1)\beta_1(n_1)\alpha_2(m_2)\beta_2(n_2)\frac{n_2}{m_1}e^{-2\pi\left(\frac{m_1 n_1}{N^2 y} + m_2 n_2 y\right)}\mathrm{d}y.$$

Integrating, we get

$$\int_{c/N}^{\infty} \eta_1 = \sum_{m_1,n_1,m_2,n_2 \geq 1} \alpha_1(m_1)\beta_1(n_1)\alpha_2(m_2)\beta_2(n_2)\frac{n_2}{m_1} \int_0^{\infty} e^{-2\pi\left(\frac{m_1 n_1}{N^2 y} + m_2 n_2 y\right)}\mathrm{d}y.$$

Now comes the trick: we make the change of variables $y \mapsto \frac{m_1}{n_2}y$ inside the integral. We get

$$\int_0^{\infty} e^{-2\pi\left(\frac{m_1 n_1}{N^2 y} + m_2 n_2 y\right)}\mathrm{d}y = \frac{m_1}{n_2} \int_0^{\infty} e^{-2\pi\left(\frac{n_1 n_2}{N^2 y} + m_1 m_2 y\right)}\mathrm{d}y.$$

This gives

$$\int_{c/N}^{\infty} \eta_1 = \sum_{m_1,n_1,m_2,n_2 \geq 1} \alpha_1(m_1)\beta_1(n_1)\alpha_2(m_2)\beta_2(n_2) \int_0^{\infty} e^{-2\pi\left(\frac{n_1 n_2}{N^2 y} + m_1 m_2 y\right)}\mathrm{d}y$$

$$= \int_0^{\infty} \left(\sum_{m_1,m_2 \geq 1} \alpha_1(m_1)\alpha_2(m_2)e^{-2\pi m_1 m_2 y} \right)$$

$$\times \left(\sum_{n_1,n_2 \geq 1} \beta_1(n_1)\beta_2(n_2)e^{-\frac{2\pi n_1 n_2}{N^2 y}} \right)\mathrm{d}y$$

$$= \int_0^{\infty} f(iy)g\left(\frac{i}{N^2 y}\right)\mathrm{d}y$$

with

$$f(\tau) = \sum_{m_1, m_2 \geq 1} \alpha_1(m_1) \alpha_2(m_2) q^{m_1 m_2}$$

and

$$g(\tau) = \sum_{n_1, n_2 \geq 1} \beta_1(n_1) \beta_2(n_2) q^{n_1 n_2}.$$

Something has changed! There is no power of m_i or n_i in the sums defining f and g. In fact, the functions f and g are (up to the constant terms) modular forms.

In order to state the final result taking care of the constant terms, we need additional notation and definitions. For any $a, b \in \mathbb{Z}/N\mathbb{Z}$, define the Eisenstein series

$$e_{a,b} = a_0(a, b) + \sum_{\substack{m,n \geq 1 \\ m \equiv a, \, n \equiv b \bmod N}} q^{mn/N} - \sum_{\substack{m,n \geq 1 \\ m \equiv -a, \, n \equiv -b \bmod N}} q^{mn/N} \qquad (9.4)$$

where

$$a_0(a, b) = \begin{cases} \frac{1}{2} - \{\frac{b}{N}\} & \text{if } a = 0, \ b \neq 0, \\ \frac{1}{2} - \{\frac{a}{N}\} & \text{if } a \neq 0, \ b = 0, \\ 0 & \text{otherwise,} \end{cases}$$

and $\{x\} = x - \lfloor x \rfloor$ is the fractional part of x. For any $a, b, c \in \mathbb{Z}/N\mathbb{Z}$, define

$$f_{a,b,c} = e_{a,bc} e_{ac,-b} + e_{a,-bc} e_{ac,b}$$

as a power series in $q^{1/N}$ with rational coefficients.

Exercise 9.7 (a) Show that $e_{a,b}$ is a linear combination of the $E^*_{1,(c,d)}$ defined in Exercise 9.12(c), hence belongs to $M_1(\Gamma(N))$.
(b) Show that $f_{a,b,c} \in M_2(\Gamma_1(N)) \cap \mathbb{Q}[[q]]$.

In our case $f_{a,b,c}$ is a modular form of weight 2, so that $L(f_{a,b,c}, s)$ converges for $\mathrm{Re}(s) > 2$. This can also be proved directly from the definition of $f_{a,b,c}$.

Theorem 9.8 ([223]) *Let $N \geq 1$ be an integer, $a, b \in \{1, \ldots, N - 1\}$ and $c \in \mathbb{Z}$. Then*

$$\int_{c/N}^{\infty} \eta(\tilde{g}_a, \tilde{g}_b) = \pi \Lambda^*(f_{a,b,c}, 0).$$

We may now finish the proof of Theorem 9.1. Going back to (9.3) and using Theorem 9.8, we get $m(P) = \Lambda^*(f, 0)$ with

$$f = f_{8,1,3} - f_{8,11,3} - f_{2,1,3} + f_{2,11,3}.$$

By Exercise 9.7(b), the function f is a modular form of weight 2 on $\Gamma_1(15)$. We can compute the coefficients of the power series f, in particular, the constant term is 0, and thus $\Lambda^*(f, 0) = \Lambda(f, 0) = L'(f, 0)$.

It remains to identify the modular form f and to relate it to $E \cong X_1(15)$. Since $X_1(15)$ has genus 1, the space $S_2(\Gamma_1(15))$ is one-dimensional, spanned by a cusp form $f_{15} = \sum_{n \geq 1} a_n q^n$, normalised by the condition $a_1 = 1$. We may compute the Fourier expansion of f_{15} by noting that $f_{15}(\tau) \, d\tau \in \Omega^1(X_1(15))$ is proportional to $\varphi^*(\omega_E)$ where $\omega_E = du/(u(v - 1/v))$ is an invariant differential on E. By computing the Fourier coefficients up to the Sturm bound, we find in fact $f = f_{15}$. Now the theory of Eichler and Shimura tells us that $L(f_{15}, s)$ coincides with $L(E, s)$, because the elliptic curve associated to f_{15} is simply $X_1(15) \cong E$. It follows that

$$\mathrm{m}(P) = L'(f, 0) = L'(f_{15}, 0) = L'(E, 0).$$

This concludes the proof of Theorem 9.1.   □

Chapter notes

More generally, given a curve $C : P(x, y) = 0$, evaluating $\mathrm{m}(P)$ with the Rogers–Zudilin method will work provided the following two conditions hold:

(1) The curve C can be parametrised by modular units $(x(\tau), y(\tau))$.
(2) The Deninger path γ_P on C is homologous to the push-forward of a modular symbol.

A way to fulfil condition (1) is to take any two modular units u and v and look at their minimal polynomial $P_{u,v} \in \mathbb{Q}[x, y]$.

One can show [50] that there are only *finitely many* elliptic curves over $\mathbb{Q}$ satisfying (1). Condition (2) is not always satisfied, and it is not clear when it will hold. Although not needed for the method to work, it would be interesting to know, in the case that C is parametrised by a modular curve, under which conditions γ_P is *equal* to the push-forward of a geodesic in $\mathcal{H} \cup \mathbb{P}^1(\mathbb{Q})$.

Although the two conditions above are not satisfied in general, K-theoretical considerations are helpful. For example, let $E : P(x, y) = 0$ be an elliptic curve defined over $\mathbb{Q}$ and let $\varphi \colon X_1(N) \to E$ be its modular parametrisation. One may try to express the pull-back $\varphi^*\{x, y\}$ as a linear combination of cup-products of modular units. This is not always possible since this pull-back lives in $K_2(\mathbb{Q}(X_1(N)))$, while the cup-products of modular units live in $K_2(Y_1(N))$, a much smaller space, as witnessed by Quillen's exact localisation sequence (7.5).

However, if $P \in \mathbb{Q}[x, y]$ is *tempered* then $\{x, y\}$ extends to $K_2(E) \otimes \mathbb{Q}$ and thus $\varphi^*\{x, y\} \in K_2(X_1(N)) \otimes \mathbb{Q}$. The question is then whether $K_2(X_1(N))$ is generated by cup-products of Siegel units of level N. Conditionally on Beilinson's conjecture, this is true if N is prime [49, Theorem 4.4]; for general N, and still conditionally, Beilinson's theorem implies that $K_2(X_1(N))$ and even $K_2(X(N))$ are generated by cup-products of Siegel units of some unspecified level divisible by N [14, Theorem 5.1.2], [174].

If, moreover, P does not vanish on the torus $\mathbb{T}^2$, then the path γ_P is closed on E. In this case, some multiple of γ_P will be homologous to the push-forward of a *closed* path $\tilde{\gamma}$ on $X_1(N)$, and Manin's theorem [133, Proposition 1.6] allows us to write $\tilde{\gamma}$ in terms of modular symbols, so that we are in the scope of the Rogers–Zudilin method. In general however, γ_P is not closed and its boundary may consist of points that are not cusps.

Additional exercises

Exercise 9.8 (Siegel units) Let $a, b \in \mathbb{Z}/N\mathbb{Z}$, $(a, b) \neq (0, 0)$. The Siegel unit $g_{a,b}$ is defined by

$$g_{a,b}(\tau) = q^{B_2(\tilde{a}/N)/2} \prod_{n \geq 0} (1 - q^{n + \tilde{a}/N} \zeta_N^b) \prod_{n \geq 1} (1 - q^{n - \tilde{a}/N} \zeta_N^{-b}) \quad \text{for } \tau \in \mathcal{H},$$

where $\tilde{a} \in \{0, \ldots, N-1\}$ is the lift of a, $B_2(x) = x^2 - x + \frac{1}{6}$, $q^\lambda = \exp(2\pi i \lambda \tau)$ and $\zeta_N = \exp(2\pi i/N)$. The aim of the exercise is to demonstrate the transformation properties of $g_{a,b}$ with respect to $\mathrm{SL}_2(\mathbb{Z})$ showing, in particular, when products or quotients of Siegel units are modular.

We first recall the definitions of the Weierstrass functions $\wp$ and ζ. Let $\tau \in \mathcal{H}$ and $\Lambda_\tau = \mathbb{Z}\tau + \mathbb{Z}$. For any $z \in \mathbb{C}$, we define

$$\wp(z, \tau) = \frac{1}{z^2} + \sum_{\substack{\omega \in \Lambda_\tau \\ \omega \neq 0}} \frac{1}{(z - \omega)^2} - \frac{1}{\omega^2},$$

$$\zeta(z, \tau) = \frac{1}{z} + \sum_{\substack{\omega \in \Lambda_\tau \\ \omega \neq 0}} \frac{1}{z - \omega} + \frac{1}{\omega} + \frac{z}{\omega^2}.$$

(a) Show that $\frac{d}{dz}\zeta(z, \tau) = -\wp(z, \tau)$.

(b) Show that there exists a linear map $\phi(\,\cdot\,, \tau) \colon \Lambda_\tau \to \mathbb{C}$ such that $\zeta(z + \omega, \tau) = \zeta(z, \tau) + \phi(\omega, \tau)$ for any $\omega \in \Lambda_\tau$. The map ϕ is the *quasi-period map*, also known as the Weierstrass eta function.

(c) Show the Legendre relation $\tau\phi(1, \tau) - \phi(\tau, \tau) = 2\pi i$.

(d) Show that for any matrix $\left(\begin{smallmatrix} a & b \\ c & d \end{smallmatrix}\right) \in \mathrm{SL}_2(\mathbb{Z})$,

$$\phi(1, \gamma\tau) = (c\tau + d)^2 \phi(1, \tau) - 2\pi i c(c\tau + d).$$

The Weierstrass function σ is defined by

$$\sigma(z, \tau) = z \prod_{\substack{\omega \in \Lambda_\tau \\ \omega \neq 0}} \left(1 - \frac{z}{\omega}\right) \exp\left(\frac{z}{\omega} + \frac{1}{2}\left(\frac{z}{\omega}\right)^2\right).$$

It can be expressed as the following infinite product (see [192]):

$$\sigma(z, \tau) = -\frac{1}{2\pi i} e^{\frac{1}{2}\phi(1,\tau)z^2} e^{-\pi i z} \frac{\prod_{n \geq 0}(1 - q^n x) \prod_{n \geq 1}(1 - q^n x^{-1})}{\prod_{n \geq 1}(1 - q^n)^2},$$

where $x = e^{2\pi i z}$ and $q = e^{2\pi i \tau}$. Also recall definition (6.3) of the Dedekind eta function $\eta(\tau)$.

(e) Using the fact that $\Delta = \eta^{24}$ is a modular form of weight 12 on $\mathrm{SL}_2(\mathbb{Z})$, show that for any $\gamma = \left(\begin{smallmatrix} a & b \\ c & d \end{smallmatrix}\right) \in \mathrm{SL}_2(\mathbb{Z})$, there exists a 12th root of unity ζ such that $\eta(\gamma\tau)^2 = \zeta \times (c\tau + d)\eta(\tau)^2$.

(f) Express the Siegel unit $g_{a,b}$ in terms of σ and η.

(g) Show that for every $\gamma \in \mathrm{SL}_2(\mathbb{Z})$, there exists a $(12N^2)$th root of unity ζ such that $g_{a,b} \circ \gamma = \zeta \times g_{(a,b)\gamma}$.

(h) Let $\Gamma(N) = \ker(\mathrm{SL}_2(\mathbb{Z}) \to \mathrm{SL}_2(\mathbb{Z}/N\mathbb{Z}))$. Show that $g_{a,b}$ is modular for $\Gamma(12N^2)$ and that $g_{a,b}^{12N}$ is modular for $\Gamma(N)$.

(i) Let $(e_{a,b})$ be a family of integers indexed by $(\mathbb{Z}/N\mathbb{Z})^2 \setminus \{0\}$ such that

$$\sum_{a,b} e_{a,b} \equiv 0 \bmod 12, \quad \sum_{a,b} a^2 e_{a,b} \equiv \sum_{a,b} ab e_{a,b} \equiv \sum_{a,b} b^2 e_{a,b} \equiv 0 \bmod 2N.$$

Show that $u = \prod_{a,b} g_{a,b}^{e_{a,b}}$ is modular for $\Gamma(N)$.

Hint (c) Integrate ζ on a suitable fundamental domain of $\mathbb{C}/\Lambda_\tau$. □

Exercise 9.9 The aim of this exercise is to construct modular units on $Y_1(N)$. Recall the function $\tilde{g}_c$ from Section 9.1 defined by

$$\tilde{g}_c(\tau) = q^{NB_2(c/N)/2} \prod_{\substack{n \geq 1 \\ n \equiv c \bmod N}} (1 - q^n) \prod_{\substack{n \geq 1 \\ n \equiv -c \bmod N}} (1 - q^n), \quad \text{where } \tau \in \mathcal{H},$$

for $1 \leq c \leq N - 1$, using the same notation as in Exercise 9.8.

(a) Express $\tilde{g}_c$ in terms of Siegel units.

(b) Using Exercise 9.8, show that $\tilde{g}_c^{12N} \in O(Y_1(N))^\times$.

(c) Let (e_c) be a family of integers indexed by $c \in \{1, \ldots, N-1\}$ such that

$$\sum_c e_c \equiv 0 \bmod 12, \quad \sum_c c e_c \equiv 0 \bmod 2, \quad \sum_c c^2 e_c \equiv 0 \bmod 2N.$$

Show that $u = \prod_c \tilde{g}_c^{e_c}$ is modular for $\Gamma_1(N)$.

Hint (b) and (c) Use the explicit transformation formulae for Siegel units proved in the course of Exercise 9.8. □

Exercise 9.10 The aim of this exercise is to give some sufficient conditions for the differential form $\eta(\tilde{g}_c, \tilde{g}_d)$ to have trivial residues at cusps.

(a) Show that for any $c, d \in \{1, \ldots, N-1\}$, we have $\mathrm{Res}_\infty \eta(\tilde{g}_c, \tilde{g}_d) = 0$.
(b) Using Exercise 9.8, show that for any $\gamma \in \mathrm{SL}_2(\mathbb{Z})$, we have $\log|g_{a,b} \circ \gamma| = \log|g_{(a,b)\gamma}|$.
(c) Show that for any $c, d \in \{1, \ldots, N-1\}$ prime to N and any cusp $\alpha = p/q$ with $(p, N) = 1$ and $(q, N) > 1$, we have $\mathrm{Res}_\alpha(\eta(\tilde{g}_c, \tilde{g}_d)) = 0$.
(d) Show that for any $c, d \in \{1, \ldots, N-1\}$ and any cusp $\alpha = p/N$ with pc and pd not divisible by N, we have $\mathrm{Res}_\alpha(\eta(\tilde{g}_c, \tilde{g}_d)) = 0$.

Exercise 9.11 Kronecker's second limit formula [191] states that

$$\lim_{s \to 1} \sum_{\substack{(m,n) \in \mathbb{Z}^2 \\ (m,n) \neq (0,0)}} \frac{\zeta_N^{ma-nb}}{|m + n\tau|^{2s}} = -\frac{2\pi}{\mathrm{Im}(\tau)} \log|g_{a,b}(\tau)|.$$

Use this formula to give another proof of the identity $\log|g_{a,b} \circ \gamma| = \log|g_{(a,b)\gamma}|$ for $(a, b) \in (\mathbb{Z}/N\mathbb{Z})^2 \setminus \{(0,0)\}$ and $\gamma \in \mathrm{SL}_2(\mathbb{Z})$.

Exercise 9.12 (a) Let $k \geq 4$ be an even integer. Show that the Eisenstein series

$$E_k(\tau) = \sum_{\substack{(m,n) \in \mathbb{Z}^2 \\ (m,n) \neq (0,0)}} \frac{1}{(m\tau + n)^k}$$

is well defined and belongs to $M_k(\mathrm{SL}_2(\mathbb{Z}))$.

(b) Using the Poisson summation formula, determine the Fourier expansion of E_k.
(c) (*More difficult*) Let $(a, b) \in (\mathbb{Z}/N\mathbb{Z})^2$. For $\mathrm{Re}(s) > 1$, define

$$E_{1,(a,b)}(\tau, s) = \sum_{\substack{(m,n) \equiv (a,b) \bmod N \\ (m,n) \neq (0,0)}} \frac{1}{(m\tau + n)|m\tau + n|^s}.$$

Show that $s \mapsto E_{1,(a,b)}(\tau, s)$ has analytic continuation to $\mathbb{C}$ and is holomorphic at $s = 0$.

(d) Show that $E^*_{1,(a,b)}(\tau) := E_{1,(a,b)}(\tau, 0) \in M_1(\Gamma(N))$ and compute its Fourier expansion.

Exercise 9.13 We establish here the meromorphic continuation and functional equation of the L-function attached to a modular form. Let $f(\tau) = \sum_{n=0}^{\infty} a_n q^{n/N}$ be a modular form in $M_k(\Gamma(N))$ with $k \geq 1$ and $N \geq 1$. We recall that the Fourier coefficients a_n grow at most polynomially: $|a_n| \ll n^c$ for some real constant c. Let $L(f, s) = \sum_{n=1}^{\infty} a_n \left(\frac{n}{N}\right)^{-s}$ be the L-function of f.

(a) Show that for $\operatorname{Re}(s) > c + 1$, we have

$$(2\pi)^{-s}\Gamma(s)L(f, s) = \int_0^{\infty} (f(iy) - a_0)y^s \frac{dy}{y}. \tag{9.5}$$

(b) Let $\sigma = \left(\begin{smallmatrix} 0 & -1 \\ 1 & 0 \end{smallmatrix}\right) \in \operatorname{SL}_2(\mathbb{Z})$. Recall that $(f|_k\sigma)(\tau) = \tau^{-k}f(-1/\tau)$. Show that $f|_k\sigma \in M_k(\Gamma(N))$.

(c) Show that the right-hand side of (9.5) has a meromorphic continuation to $\mathbb{C}$ and is holomorphic everywhere except possibly simple poles at $s = 0$ and $s = k$.

(d) Show that the completed L-function $\Lambda(f, s) = N^{s/2}(2\pi)^{-s}\Gamma(s)L(f, s)$ satisfies the functional equation $\Lambda(f, s) = i^k N^{s-\frac{k}{2}}\Lambda(f|_k\sigma, k - s)$.

(e) In the case $N = 1$ and $k \equiv 2 \bmod 4$, show that $L(f, k/2) = 0$.

(f) For any matrix $g = \left(\begin{smallmatrix} a & b \\ c & d \end{smallmatrix}\right) \in \operatorname{GL}_2^+(\mathbb{R})$, we define

$$(f|_k g)(\tau) = \det(g)^{k/2}(c\tau + d)^{-k}f\left(\frac{a\tau + b}{c\tau + d}\right).$$

Show that the matrix $W_N = \left(\begin{smallmatrix} 0 & -1 \\ N & 0 \end{smallmatrix}\right)$ normalises $\Gamma_1(N)$ and deduce that for every $f \in M_k(\Gamma_1(N))$, we have $f|_k W_N \in M_k(\Gamma_1(N))$.

(g) Assuming $f \in M_k(\Gamma_1(N))$, show the (cleaner) functional equation

$$\Lambda(f, s) = i^k \Lambda(f|_k W_N, k - s).$$

(h) Determine the residues of $\Lambda(f, s)$ at $s = 0$ and $s = k$.

Hint (c) Split the integral $\int_0^{\infty} = \int_0^1 + \int_1^{\infty}$. $\square$

Exercise 9.14 (Diamond operators) Let $N \geq 1$ be an integer. For any $\delta \in (\mathbb{Z}/N\mathbb{Z})^{\times}$, let $\langle\delta\rangle$ be a matrix in $\operatorname{SL}_2(\mathbb{Z})$ such that

$$\langle\delta\rangle \equiv \begin{pmatrix} \delta^{-1} & * \\ 0 & \delta \end{pmatrix} \bmod N.$$

(a) Show that $\Gamma_1(N)$ is a normal subgroup of $\Gamma_0(N)$, and that $\Gamma_0(N)/\Gamma_1(N)$ is isomorphic to $(\mathbb{Z}/N\mathbb{Z})^{\times}$.

(b) Show that $\tau \mapsto \langle\delta\rangle\tau$ induces an automorphism of $X_1(N)$ that depends only on δ and not on the particular matrix $\langle\delta\rangle$. This automorphism is called a *diamond operator*.

(c) Show that the diamond operators give rise to a faithful action of the quotient group $(\mathbb{Z}/N\mathbb{Z})^\times/\{\pm 1\}$ on $X_1(N)$.

(d) Recall that we have an isomorphism $\varphi\colon X_1(15) \xrightarrow{\cong} E$, where E is the elliptic curve defined in Exercise 9.4. Show that the map $\delta \mapsto \varphi(\langle\delta\rangle\infty)$ induces a group isomorphism between $(\mathbb{Z}/15\mathbb{Z})^\times/\{\pm 1\}$ and $E(\mathbb{Q}) \cong \mathbb{Z}/4\mathbb{Z}$.

(e) Show that the action of diamond operators on $X_1(15)$ corresponds, via the isomorphism φ, to the action of $E(\mathbb{Q})$ by translation on E.

10

Modular regulators

In Chapter 9 we defined modular curves as quotients of the upper half-plane. Another useful perspective on modular curves is to view them as moduli spaces: they classify elliptic curves (with additional structure). We define the universal elliptic curve $\mathcal{E}(N)$ in Section 10.1 and in subsequent sections investigate the Beilinson regulator map attached to $\mathcal{E}(N)$ and more generally to its fibred powers $\mathcal{E}(N)^k$. The main result (Theorem 10.11) relates these regulators to special values of L-functions of modular forms, in accordance with Beilinson's conjectures.

In order to be able to state the theorem, we need to introduce several objects. In Section 10.2, we define the Shokurov cycles, which are special topological cycles on $\mathcal{E}(N)^k$, generalising modular symbols. In Section 10.3, we define the Eisenstein symbols, which are an algebraic incarnation of Eisenstein series, and live in the motivic cohomology of $\mathcal{E}(N)^k$. In Section 10.4, we define the Beilinson–Deninger–Scholl elements, which are (roughly) cup-products of Eisenstein symbols. The Beilinson regulator is then obtained by integrating these classes over the Shokurov cycles, and can be computed using the Rogers–Zudilin method outlined in Chapter 9. In the final Section 10.5, we describe some recent applications of Theorem 10.11 to Mahler measures of elliptic surfaces.

10.1 Universal elliptic curves

The interpretation of modular curves as classifying spaces of elliptic curves can be explained as follows. To any τ in the upper half-plane $\mathcal{H}$, we may associate the elliptic curve $E_\tau := \mathbb{C}/(\mathbb{Z} + \tau\mathbb{Z})$. Since every elliptic curve over $\mathbb{C}$ is isomorphic to $\mathbb{C}/\Lambda$ for some lattice Λ in $\mathbb{C}$, and every lattice is homothetic to $\mathbb{Z} + \tau\mathbb{Z}$ for some $\tau \in \mathcal{H}$, we get a surjective map

$$\mathcal{H} \longrightarrow \{\text{isomorphism classes of elliptic curves over } \mathbb{C}\}. \qquad (10.1)$$

What are the fibres of this map? Two elliptic curves $\mathbb{C}/\Lambda$ and $\mathbb{C}/\Lambda'$ are isomorphic as complex-analytic manifolds if and only if Λ and Λ' are homothetic. Furthermore, two lattices $\mathbb{Z} + \tau\mathbb{Z}$ and $\mathbb{Z} + \tau'\mathbb{Z}$ are homothetic if and only if τ and τ' are equivalent under the action of $SL_2(\mathbb{Z})$. In other words, (10.1) induces a bijection

$$SL_2(\mathbb{Z}) \backslash \mathcal{H} \xrightarrow{\cong} \{\text{isomorphism classes of elliptic curves over } \mathbb{C}\}. \qquad (10.2)$$

This generalises to arbitrary congruence subgroups. The next exercise illustrates this in the case of $\Gamma_1(N)$.

Exercise 10.1 Let $N \geq 1$ be an integer. Show that the map $\tau \mapsto (E_\tau, 1/N)$ induces a bijection

$$Y_1(N) = \Gamma_1(N) \backslash \mathcal{H} \xrightarrow{\cong} \left\{ (E, P) : \begin{array}{l} E/\mathbb{C} \text{ is an elliptic curve} \\ P \in E \text{ has order } N \end{array} \right\}/_\sim \qquad (10.3)$$

where $(E, P) \sim (E', P')$ if and only if there exists an isomorphism of elliptic curves $E \xrightarrow{\cong} E'$ sending P to P'.

Via the isomorphisms above, the canonical projection

$$\Gamma_1(N) \backslash \mathcal{H} \to SL_2(\mathbb{Z}) \backslash \mathcal{H}$$

is given by 'forgetting the additional structure', namely $(E, P) \mapsto E$.

Since modular curves classify all elliptic curves, we expect the existence of a universal object: given a modular curve Y, there should be some space $\mathcal{E}_Y$ lying over Y with the property that the fibre $\mathcal{E}_y$ over $y \in Y$ is precisely the elliptic curve corresponding to y. Such an object is called a *universal elliptic curve*. It is not difficult to construct such an object over $\mathcal{H}$. Namely, consider the quotient $\mathcal{E} := \mathbb{Z}^2 \backslash (\mathcal{H} \times \mathbb{C})$, where $\mathbb{Z}^2$ acts on $\mathcal{H} \times \mathbb{C}$ as follows:

$$(m, n) \cdot (\tau; z) := (\tau; z + m\tau + n) \quad \text{for } (m, n) \in \mathbb{Z}^2, \ \tau \in \mathcal{H}, \ z \in \mathbb{C}. \qquad (10.4)$$

In other words, if we view $\mathcal{H} \times \mathbb{C}$ as a space over $\mathcal{H}$, then $\mathbb{Z}^2$ acts fibrewise and over $\tau \in \mathcal{H}$, the corresponding quotient is the elliptic curve E_τ. The space $\mathcal{E}$ is a complex-analytic manifold of dimension 2 and by construction, there is a holomorphic projection map $\mathcal{E} \to \mathcal{H}$ with the universal property stated above.

In order to define universal elliptic curves over arbitrary modular curves, we want to take the quotient of $\mathcal{E}$ by congruence subgroups of $SL_2(\mathbb{Z})$. First, we need to lift the action of $SL_2(\mathbb{Z})$ on $\mathcal{H}$ to our space $\mathcal{E}$. This can be done by defining

$$\begin{pmatrix} a & b \\ c & d \end{pmatrix} \cdot (\tau; z) := \left(\frac{a\tau + b}{c\tau + d}; \frac{z}{c\tau + d} \right) \quad \text{for } \begin{pmatrix} a & b \\ c & d \end{pmatrix} \in SL_2(\mathbb{Z}). \tag{10.5}$$

This gives an action of $SL_2(\mathbb{Z})$ on $\mathcal{H} \times \mathbb{C}$, which lifts that on $\mathcal{H}$, and induces an action of $SL_2(\mathbb{Z})$ on $\mathcal{E}$.

Now, we cannot form the quotient $SL_2(\mathbb{Z}) \setminus \mathcal{E}$ in the category of complex-analytic varieties, because $SL_2(\mathbb{Z})$ does not act freely: for example, the matrix $-I_2$ acts as $(\tau; z) \mapsto (\tau; -z)$, thus fixes the two-torsion points in every fibre of $\mathcal{E}$. Similarly, the matrices $\begin{pmatrix} 0 & -1 \\ 1 & 0 \end{pmatrix}$ and $\begin{pmatrix} 0 & -1 \\ 1 & -1 \end{pmatrix}$ fix respectively the points $(i, 0)$ and $(e^{\pi i/3}, 0)$ on $\mathcal{E}$. Note that all these matrices have finite order in $SL_2(\mathbb{Z})$.

The problem disappears if we restrict the action to a *torsion-free* subgroup Γ of $SL_2(\mathbb{Z})$: such a group Γ acts freely on $\mathcal{E}$, since it already acts freely on $\mathcal{H}$ [185, p. 129, Théorème 1].

Definition 10.1 Let Γ be a congruence subgroup of $SL_2(\mathbb{Z})$ acting freely on $\mathcal{H}$, and let $Y(\Gamma) = \Gamma \setminus \mathcal{H}$ be the associated modular curve. The *universal elliptic curve* $\mathcal{E}(\Gamma)$ over $Y(\Gamma)$ is the quotient $\Gamma \setminus \mathcal{E}$.

The fibres of the canonical map $\mathcal{E}(\Gamma) \to Y(\Gamma)$ are still elliptic curves: for any $\tau \in \mathcal{H}$, the fibre of $\mathcal{E}(\Gamma)$ over $[\tau]$ is the elliptic curve E_τ.

The description of modular curves as moduli spaces is completely algebraic and makes sense over arbitrary fields of characteristic 0, for example over $\mathbb{Q}$. As a consequence, $Y(\Gamma)$ and $\mathcal{E}(\Gamma)$ are algebraic varieties. It turns out that they have canonical models defined over cyclotomic fields [187, 6.7].

Example 10.2 Here are some examples of universal elliptic curves.

(1) The group $\Gamma_1(N)$ acts freely on $\mathcal{H}$ if and only if $N \geq 4$. In this case, we denote by $\mathcal{E}_1(N) = \mathcal{E}(\Gamma_1(N))$ the associated universal elliptic curve over $Y_1(N)$.

(2) The group $\Gamma(N)$ acts freely on $\mathcal{H}$ if and only if $N \geq 3$. In this case, we denote by $\mathcal{E}(N) = \mathcal{E}(\Gamma(N))$ the associated universal elliptic curve over $Y(N)$.

The fact that $Y_1(N)$ parametrises elliptic curves endowed with a point of order N gives an additional structure on $\mathcal{E}_1(N)$, namely a section $P \colon Y_1(N) \to \mathcal{E}_1(N)$, which has order N in every fibre. Similarly, the universal elliptic curve $\mathcal{E}(N)$ is endowed with a pair of sections $P_1, P_2 \colon Y(N) \to \mathcal{E}(N)$ which form a basis of the N-torsion subgroup in every fibre. The following exercise describes these sections explicitly.

Exercise 10.2 (a) Let $N \geq 4$ be an integer. Show that the map $\tau \mapsto (\tau; 1/N)$ induces a section $P: Y_1(N) \to \mathcal{E}_1(N)$.

(b) Let $N \geq 3$ be an integer. Show that the maps $\tau \mapsto (\tau; \tau/N)$ and $\tau \mapsto (\tau; 1/N)$ induce sections $P_1, P_2: Y(N) \to \mathcal{E}(N)$.

There is a canonical degeneracy map $\mathcal{E}(N) \to \mathcal{E}_1(N)$ compatible with the natural map $Y(N) \to Y_1(N)$. The action of $SL_2(\mathbb{Z})$ on $\mathcal{E}$ and $\mathcal{H}$ induces an action of $SL_2(\mathbb{Z}/N\mathbb{Z})$ on $\mathcal{E}(N)$ and $Y(N)$.

Remark 10.3 In the literature, $Y(N)$ usually denotes the algebraic curve over $\mathbb{Q}$ which is the solution of an appropriate moduli problem [111, Section 1]. This curve is not geometrically connected, and $\Gamma(N) \backslash \mathcal{H}$ corresponds to one connected component of its complex points $Y(N)(\mathbb{C})$ [111, Section 1.8].

Notation 10.4 Let $N \geq 3$ be an integer. For any integer $k \geq 0$, we denote by $\mathcal{E}(N)^k$ the k-fold fibred product of $\mathcal{E}(N)$ over $Y(N)$.

In other words, the fibre of $\mathcal{E}(N)^k$ over the point $[\tau] \in Y(N)$ is the kth power E_τ^k of the elliptic curve E_τ. A point of $\mathcal{E}(N)^k$ can be represented by a tuple $(\tau; z_1, \ldots, z_k)$ with $\tau \in \mathcal{H}$ and $z_1, \ldots, z_k \in \mathbb{C}$. Note that $\mathcal{E}(N)^k$ is a complex-analytic manifold of dimension $k + 1$. Alternatively, $\mathcal{E}(N)^k$ can be described as the quotient of $\mathcal{H} \times \mathbb{C}^k$ by the semidirect product $(\mathbb{Z}^2)^k \rtimes \Gamma(N)$, where we view the elements of $\mathbb{Z}^2$ as row vectors, and $\gamma \in \Gamma(N)$ acts on $(\mathbb{Z}^2)^k$ by $(v_i) \mapsto (v_i \gamma^{-1})$.

Exercise 10.3 Show that $M_{k+2}(\Gamma(N))$ embeds into the space of holomorphic differential $(k + 1)$-forms on $\mathcal{E}(N)^k$, by sending a modular form f to

$$\omega_f = (2\pi i)^{k+1} f(\tau) \, d\tau \wedge dz_1 \wedge \cdots \wedge dz_k.$$

One can show that the image of $M_{k+2}(\Gamma(N))$ is contained in the space of holomorphic differential forms with logarithmic singularities at infinity.

10.2 Shokurov cycles

The Shokurov cycles are topological $(k + 1)$-chains on $\mathcal{E}(N)^k$ generalising the modular symbols on the modular curve $Y(N)$. More precisely, the Shokurov cycles are fibred over modular symbols. Their fibres are k-cycles on E_τ^k which can be explicitly described.

The homology $H_1(E_\tau, \mathbb{Z})$ of the elliptic curve $E_\tau = \mathbb{C}/(\mathbb{Z} + \tau\mathbb{Z})$ is a free abelian group of rank 2. A basis of $H_1(E_\tau, \mathbb{Z})$ is given by the cycles $\gamma_{0 \to \tau}$ and $\gamma_{0 \to 1}$ depicted in Figure 10.1. The path $\gamma_{0 \to \tau}$ (respectively, $\gamma_{0 \to 1}$) is the image of the segment $[0, \tau]$ (respectively, $[0, 1]$) under the canonical projection map $\mathbb{C} \to E_\tau$.

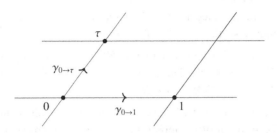

Figure 10.1 Basis of $H_1(E_\tau, \mathbb{Z})$

Taking cross products [99, 3.B], we get homology classes $\gamma_1 \times \cdots \times \gamma_k$ in the homology group $H_k(E_\tau^k, \mathbb{Z})$ for any choice of $\gamma_1, \ldots, \gamma_k \in H_1(E_\tau, \mathbb{Z})$.

Definition 10.5 Let $k \geq 0$ be an integer. Let $\alpha, \beta \in \mathbb{P}^1(\mathbb{Q})$ and let j be an integer satisfying $0 \leq j \leq k$. The *Shokurov cycle* $X^j Y^{k-j}\{\alpha, \beta\}$ on $\mathcal{E}(N)^k$ is the $(k+1)$-chain fibred over the modular symbol $\{\alpha, \beta\}$ in $Y(N)$ whose fibre at every $\tau \in \{\alpha, \beta\}$ is given by $\gamma_{0 \to \tau}^j \times \gamma_{0 \to 1}^{k-j}$.

Explicitly, we have

$$X^j Y^{k-j}\{\alpha, \beta\} = \{(\tau; t_1\tau, \ldots, t_j\tau, t_{j+1}, \ldots, t_k) : \tau \in \{\alpha, \beta\}, t_1, \ldots, t_k \in [0, 1]\}.$$
(10.6)

We endow $X^j Y^{k-j}\{\alpha, \beta\}$ with the orientation induced by the product orientation on $\{\alpha, \beta\} \times [0, 1]^k$. Note that for $k = 0$, the Shokurov cycle above is just the usual modular symbol $\{\alpha, \beta\}$.

10.3 Eisenstein symbols

The Eisenstein symbols are distinguished cohomology classes on the (fibred) powers of the universal elliptic curve. In Deligne–Beilinson cohomology, they are described by certain explicit real analytic Eisenstein series. In fact, Beilinson defined them at the motivic level, and showed that their images under the regulator map are given by the above Eisenstein series. References for the material in this section include [15, 70, 69, 103].

From now on, we will view the modular curve $Y(N)$ and its compactification $X(N)$ as algebraic curves defined over $\mathbb{Q}$ (see Remark 10.3). The finite set of closed points $C_N = X(N) \setminus Y(N)$ is the set of *cusps* of the modular curve. There are bijections

$$C_N(\mathbb{C}) \cong (\mathbb{Z}/N\mathbb{Z})^\times \times (\Gamma(N) \setminus \mathbb{P}^1(\mathbb{Q})) \quad \text{and} \quad C_N \cong (\pm P) \setminus \mathrm{GL}_2(\mathbb{Z}/N\mathbb{Z}),$$

where P denotes the subgroup $\{\left(\begin{smallmatrix} * & * \\ 0 & 1 \end{smallmatrix}\right)\}$ of $\mathrm{GL}_2(\mathbb{Z}/N\mathbb{Z})$.

Accordingly, the group $H^1_{\mathcal{M}}(Y(N), \mathbb{Q}(1)) \cong O(Y(N))^\times \otimes \mathbb{Q}$ is the group of modular units *defined over* $\mathbb{Q}$. We have the divisor map

$$\mathrm{Res}^0 \colon H^1_{\mathcal{M}}(Y(N), \mathbb{Q}(1)) \to \mathbb{Q}[C_N]^0,$$

where $\mathbb{Q}[C_N]^0$ is the $\mathbb{Q}$-vector space of divisors of degree 0 on C_N.

For an integer $N \geq 3$, we denote by $E(N)$ the universal elliptic curve over $Y(N)$. It is an algebraic variety defined over $\mathbb{Q}$. The complex-analytic manifold $\mathcal{E}(N)$ introduced in Section 10.1 is isomorphic to one connected component of $E(N)(\mathbb{C})$. The divisor map Res^0 introduced above can be generalised to the powers of $E(N)$, namely, for every $k \geq 1$ there is a *residue map*

$$\mathrm{Res}^k \colon H^{k+1}_{\mathcal{M}}(E(N)^k, \mathbb{Q}(k+1)) \to \mathbb{Q}[C_N]^{(k)},$$

where

$$\mathbb{Q}[C_N]^{(k)} = \{f \colon P \setminus \mathrm{GL}_2(\mathbb{Z}/N\mathbb{Z}) \to \mathbb{Q} \colon f(-g) = (-1)^k f(g) \text{ for all } g\}.$$

The vector space $\mathbb{Q}[C_N]^{(k)}$ is non-canonically isomorphic to $\mathbb{Q}[C_N]$. By convention, we put $\mathbb{Q}[C_N]^{(0)} := \mathbb{Q}[C_N]^0$.

Theorem 10.6 ([15, Section 3], [69]) *Let $k \geq 0$ be an integer. There is a natural map*

$$\mho^k_{\mathcal{M}} \colon \mathbb{Q}[C_N]^{(k)} \to H^{k+1}_{\mathcal{M}}(E(N)^k, \mathbb{Q}(k+1))$$

which is a right inverse of Res^k, *namely* $\mathrm{Res}^k \circ \mho^k_{\mathcal{M}} = \mathrm{id}$.

This shows, in particular, that the residue map Res^k is surjective. In the case $k = 0$, this is the famous Manin–Drinfeld theorem [133, 78]. For $k \geq 1$, Beilinson's construction of $\mho^k_{\mathcal{M}}$ for $k \geq 1$ is completely explicit and involves cup-products of (universal) elliptic functions whose zeros and poles are N-torsion sections of $E(N)$.

Definition 10.7 Let $k \geq 0$ be an integer and $u \in (\mathbb{Z}/N\mathbb{Z})^2$, with $u \neq (0,0)$ in the case $k = 0$. The *Eisenstein symbol* $\mathrm{Eis}^k(u)$ is defined by

$$\mathrm{Eis}^k(u) = \mho^k_{\mathcal{M}}(f_{k,u}) \in H^{k+1}_{\mathcal{M}}(E(N)^k, \mathbb{Q}(k+1)),$$

where $f_{k,u} \in \mathbb{Q}[C_N]^{(k)}$ is defined by $f_{k,u}(g) = B_{k+2}(\{\frac{(gu)_2}{N}\})$, where B_{k+2} is the Bernoulli polynomial (see Exercise 10.7).

In the case $k = 0$, we have $\mathrm{Eis}^0(u) = g_u \otimes (2/N)$ where g_u is the Siegel unit introduced in Exercise 9.8.

The Beilinson regulator of $\mathrm{Eis}^k(u)$, which lives in $H_{\mathcal{D}}^{k+1}(E(N)_{\mathbb{R}}^k, \mathbb{R}(k+1))$, is represented by a real analytic differential k-form $\mathrm{Eis}_{\mathcal{D}}^k(u)$ on $E(N)^k(\mathbb{C})$ satisfying $\mathrm{d}\,\mathrm{Eis}_{\mathcal{D}}^k(u) = \pi_k(\mathrm{Eis}_{\mathrm{hol}}^k(u))$, with the notation $\pi_k(\omega) = \frac{1}{2}(\omega + (-1)^k \overline{\omega})$, and where $\mathrm{Eis}_{\mathrm{hol}}^k(u)$ is a holomorphic Eisenstein series of weight $k + 2$ on $\Gamma(N)$. This explains the terminology 'Eisenstein symbol'.

Although the Eisenstein symbol is related to Eisenstein series, it can be used to construct cohomology classes related to *cusp forms*. The (rough) idea is to consider the cup-product of two Eisenstein symbols. This is similar to the work of Borisov and Gunnells [30] and Khuri-Makdisi [112], who constructed holomorphic cusp forms out of products of Eiseinstein series.

More precisely, let $k \geq 0$ be an integer, and choose a decomposition $k = k_1 + k_2$ with $k_1, k_2 \geq 0$. Consider the two canonical projections

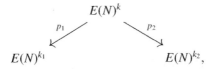

where p_1 (respectively, p_2) is defined using the first k_1 (respectively, last k_2) factors of $E(N)^k$.

Definition 10.8 Let $u_1, u_2 \in (\mathbb{Z}/N\mathbb{Z})^2$, with $u_i \neq (0,0)$ if $k_i = 0$. The *Beilinson–Deninger–Scholl elements* are defined by

$$\mathrm{Eis}^{k_1, k_2}(u_1, u_2) = p_1^* \mathrm{Eis}^{k_1}(u_1) \cup p_2^* \mathrm{Eis}^{k_2}(u_2) \in H_{\mathcal{M}}^{k+2}(E(N)^k, \mathbb{Q}(k+2)).$$

Example 10.9 For $k_1 = k_2 = 0$, we have $\mathrm{Eis}^{0,0}(u_1, u_2) = \{g_{u_1}, g_{u_2}\}$ in the group $K_2(Y(N)) \otimes \mathbb{Q}$, so that we recover the cup-products of modular units considered in Chapter 9.

According to Beilinson's conjecture, the regulator of $\mathrm{Eis}^{k_1, k_2}(u_1, u_2)$ should be related to the cohomology of $E(N)^k$ in degree $k + 1$. In the next section we show this is the case by computing explicitly the integral of $\mathrm{Eis}^{k_1, k_2}(u_1, u_2)$ along certain Shokurov cycles.

Remark 10.10 The variety $E(N)^k$ is not projective, but Deligne constructed a canonical smooth compactification $\overline{E(N)^k}$ of $E(N)^k$. If f is a newform of weight $k + 2$ on $\Gamma_1(N)$ then $L(f, s)$ is a factor of the L-function of $H^{k+1}(\overline{E(N)^k})$ [64].

10.4 Regulator integrals

In this section we give a brief overview of the computation of the regulator of the Beilinson–Deninger–Scholl elements. For more details, we refer the reader to [52].

We know that the regulator of $\mathrm{Eis}^{k_1,k_2}(u_1, u_2)$ is represented by the differential form

$$\mathrm{Eis}_{\mathcal{D}}^{k_1,k_2}(u_1, u_2) := p_1^* \, \mathrm{Eis}_{\mathcal{D}}^{k_1}(u_1) \cup p_2^* \, \mathrm{Eis}_{\mathcal{D}}^{k_2}(u_2).$$

For dimension reasons, the differential form $\mathrm{Eis}_{\mathcal{D}}^{k_1,k_2}(u_1, u_2)$ is closed. We may thus integrate it along Shokurov cycles.

In the higher weight case $k \geq 1$, it is (unfortunately) not always true that the integral

$$\int_{X^j Y^{k-j}\{\alpha,\beta\}} \mathrm{Eis}_{\mathcal{D}}^{k_1,k_2}(u_1, u_2)$$

is absolutely convergent. However, following [52] it is possible to define a regularised integral

$$\int_{X^j Y^{k-j}\{\alpha,\beta\}}^{*} \mathrm{Eis}_{\mathcal{D}}^{k_1,k_2}(u_1, u_2).$$

This modular integral can be computed, at least for certain Shokurov cycles. In order to state the result, define for $k \geq 1$ and $a, b \in \mathbb{Z}/N\mathbb{Z}$ the Eisenstein series

$$G_{a,b}^{(k)}(\tau) = a_0 + \sum_{\substack{m,n \geq 1 \\ m \equiv a,\, n \equiv b \bmod N}} m^{k-1} q^{mn/N} + (-1)^k \sum_{\substack{m,n \geq 1 \\ m \equiv -a,\, n \equiv -b \bmod N}} m^{k-1} q^{mn/N},$$

where $q^\lambda = e^{2\pi i \lambda \tau}$ as usual, and the sums are taken over the pairs (m, n) congruent to (a, b) (respectively, $(-a, -b)$) mod N. Then $G_{a,b}^{(k)}$ is a holomorphic Eisenstein series of weight k on $\Gamma(N)$, except in the case $k = 2$ and $a = 0$ where it is only quasi-modular. In fact, for fixed k the $G_{a,b}^{(k)}$ generate the Eisenstein subspace of $M_k(\Gamma(N))$. Note that $G_{a,b}^{(1)}$ coincides with $e_{a,b}$ from (9.4).

Theorem 10.11 ([52]) *Let $u_1 = (a_1, b_1)$, $u_2 = (a_2, b_2) \in (\mathbb{Z}/N\mathbb{Z})^2$. Assume $u_i \neq (0, 0)$ (respectively, $b_i \neq 0$) in the case $k_i = 0$ (respectively, $k_i = 1$). Then*

$$\int_{X^k\{0,\infty\}}^{*} \mathrm{Eis}_{\mathcal{D}}^{k_1,k_2}(u_1, u_2) = \frac{(k_1+2)(k_2+2)}{2N^{k+2}}(2\pi)^{k+1} i^{k_1-k_2+1}$$

$$\times \Lambda^*(G_{b_2,a_1}^{(k_2+1)} G_{b_1,-a_2}^{(k_1+1)} - G_{b_2,-a_1}^{(k_2+1)} G_{b_1,a_2}^{(k_1+1)}, 0).$$

Note that the modular form appearing in the right-hand side has weight $k+2$ and rational Fourier coefficients. The appearance of the L-value at $s = 0$ is in accordance with the Beilinson conjectures stated in Section 8.2.

In the case $k_1 = k_2 = 0$, the formula simplifies to

$$\int_0^{\infty} \eta(g_{a,b}, g_{c,d}) = \pi \Lambda^* (e_{a,d} e_{b,-c} + e_{a,-d} e_{b,c}, 0) \qquad (10.7)$$

which was obtained in [51] with the same methods as [223]. Using a theorem by Khuri-Makdisi [112] it is possible to show that the modular forms $e_{a,d} e_{b,-c} + e_{a,-d} e_{b,c}$ span $M_2(\Gamma(N); \mathbb{Q})$, the space of modular forms with rational Fourier coefficients — see Exercise 10.8 for details.

To compute the integral of $\eta(g_{a,b}, g_{c,d})$ over an arbitrary modular symbol, we can use the following facts:

- By Manin's theorem [133, Proposition 1.6], every modular symbol $\{\alpha, \beta\}$ can be written (in the homology of the modular curve relative to the cusps) as a $\mathbb{Z}$-linear combination of Manin symbols $\{\gamma 0, \gamma \infty\}$ with $\gamma \in SL_2(\mathbb{Z})$.
- We have

$$\int_{\gamma 0}^{\gamma \infty} \eta(g_{u_1}, g_{u_2}) = \int_0^{\infty} \gamma^* \eta(g_{u_1}, g_{u_2}) = \int_0^{\infty} \eta(g_{u_1 \gamma}, g_{u_2 \gamma}).$$

However, one must be careful with the residues of $\eta(g_{u_1}, g_{u_2})$. Taking them into account, one can show the following result.

Proposition 10.12 *Let $\mathcal{U}$ be the $\mathbb{Q}$-subspace of $O(Y(N))^\times \otimes \mathbb{Q}$ generated by the Siegel units $g_{a,b}$. For every $u, v \in \mathcal{U}$, there exists a linear map*

$$F_{u,v} \colon H_1(Y(N)(\mathbb{C}), \mathbb{Z}) \to M_2(\Gamma(N); \mathbb{Q})$$

such that

$$\int_\gamma \eta(u, v) = \pi \Lambda^* (F_{u,v}(\gamma), 0) \quad \text{for any } \gamma \in H_1(Y(N)(\mathbb{C}), \mathbb{Z}).$$

Remark 10.13 A good setting for treating residues is to use the Borel–Serre compactification of $Y(N)(\mathbb{C})$: it consists in adding at each cusp not a single point, but rather a whole circle S^1, parametrising the 'direction' in which one approaches the cusp. Stevens [202] described the homology of this compactification in terms of modular symbols and *modular caps* (which are paths on the boundary circles). In particular, he showed how to write any modular symbol $\{\alpha, \beta\}$ in terms of Manin symbols and modular caps.

Question 10.14 Does there exist a natural trilinear map

$$F \colon H_1(Y(N)(\mathbb{C}), \mathbb{Z}) \times \mathcal{U} \times \mathcal{U} \to M_2(\Gamma(N); \mathbb{Q})$$

such that

$$\int_\gamma \eta(u, v) = \pi \Lambda^* (F(\gamma, u, v), 0) ?$$

One problem that arises is that the map $f \in M_2(\Gamma(N); \mathbb{Q}) \mapsto \Lambda^*(f, 0)$ is not injective, as the following exercise shows.

Exercise 10.4 Let M be a proper divisor of N, and let d be a divisor of N/M. Show that if $f \in S_2(\Gamma_1(N))$ is of the form $f(\tau) = g(d\tau)$ with $g \in S_2(\Gamma_1(M))$, then $\Lambda(f, 0) = \Lambda(g, 0)$.

Question 10.15 Assuming the map F exists, how does it behave with respect to Hecke operators?

10.5 Applications

As discussed in Section 8.4, one can relate the Mahler measure of the polynomial $x + 1/x + y + 1/y + z + 1/z + 2$ to the L-value of a modular form of weight 3 at $s = 3$ [53]. Using Theorem 10.11, several identities of this type can be obtained along the same lines. We stress that the polynomials appearing in [53] are not given in advance: rather, one starts from some universal elliptic curve $\mathcal{E}$, and then tries to determine a model of $\mathcal{E}$ whose Mahler measure is related to the L-function of $H^2(\mathcal{E})$ (and thus, to L-functions of modular forms of weight 3).

The natural case to consider next is that of Calabi–Yau threefolds. Papanikolas, Rogers and Samart [150] proved the following identity:

$$m\left(\left(X + \frac{1}{X}\right)\left(Y + \frac{1}{Y}\right)\left(Z + \frac{1}{Z}\right)\left(T + \frac{1}{T}\right) - 16\right) = 8L'(f_8, 0) - 28\zeta'(-2), \quad (10.8)$$

where $f_8 = \eta_2^4 \eta_4^4$ (in the notation (6.3)) is the unique normalised cusp form in $S_4(\Gamma_0(8))$. Note that unlike the results of Rodriguez Villegas and Bertin, the modular form f_8 does not have complex multiplication.

We now describe how the polynomial appearing in (10.8) is related to the square of a universal elliptic curve, and how the methods from Section 10.4 could be applied to computing the Mahler measure associated to such threefolds. Using the change of variables $(x, y) = (XY, X/Y)$ and $(z, t) = (ZT, Z/T)$, one sees that

$$m\left(\left(X + \frac{1}{X}\right)\left(Y + \frac{1}{Y}\right)\left(Z + \frac{1}{Z}\right)\left(T + \frac{1}{T}\right) - 16\right)$$
$$= m\left(\left(x + \frac{1}{x} + y + \frac{1}{y}\right)\left(z + \frac{1}{z} + t + \frac{1}{t}\right) - 16\right).$$

Let us consider the congruence subgroup $\Gamma = \Gamma_1(4) \cap \Gamma_0(8)$. It is known that the modular curve $Y = Y(\Gamma) = \Gamma \backslash \mathcal{H}$ has genus 0. By a result of Bertin and Lecacheux [21], the universal elliptic curve $E = \mathcal{E}(\Gamma)$ is given by the affine equation $x + 1/x + y + 1/y = k$. Here k is a certain modular unit on Y generating the function field of Y. The structural map $\pi \colon E \to Y$ is given by $(x, y, k) \mapsto k$. Let $P(x, y) = x + 1/x + y + 1/y$. By definition, the fibre product $E^2 = E \times_Y E$ is given by the equation $P(x, y) = P(z, t)$. The following exercise shows that the Mahler measure of the polynomial $P(x, y) - P(z, t)$ is not related to the cusp form f_8.

Exercise 10.5 Show that

$$\mathrm{m}(x + 1/x + y + 1/y + z + 1/z + t + 1/t) = \frac{7\zeta(3)}{2\pi^2}.$$

However, we can produce a more interesting polynomial as follows. Consider the normaliser N of Γ in $\mathrm{SL}_2(\mathbb{R})$. It acts on Y and thus on the function field of Y. Moreover, any element $h \in N$ must send the generator k to some other generator $k' = k \circ h$. Since the function field is equal to $\mathbb{C}(k)$, we must have $k' = (ak + b)/(ck + d)$ for some $\left(\begin{smallmatrix} a & b \\ c & d \end{smallmatrix}\right) \in \mathrm{GL}_2(\mathbb{C})$. For example, $h = \left(\begin{smallmatrix} 1 & 1/2 \\ 0 & 1 \end{smallmatrix}\right)$ normalises Γ and we find that $k' = 16/k$. Let $E^{(h)}$ be the conjugate of E by h, namely the elliptic curve whose structural morphism is the composition $E \xrightarrow{\pi} Y \xrightarrow{h} Y$. In our case, $E^{(h)}$ is simply obtained by substituting $k \to 16/k$ into the equation for E, so that $E^{(h)} \colon P(x, y) = 16/k$. Note that the fibre product $E \times_Y E^{(h)}$ is given by

$$E \times_Y E^{(h)} \colon P(x, y)P(z, t) = 16,$$

the same polynomial that appears above. The following proposition provides a link between $E^{(h)}$ and E.

Proposition 10.16 *There exists a degree-4 isogeny $\tilde{h} \colon E \to E^{(h)}$ of elliptic curves over Y.*

Proof Taking the complex points and working fibrewise, we have to construct an isogeny $\mathbb{C}/\langle 1, \tau \rangle \to \mathbb{C}/\langle 1, \tau + 1/2 \rangle$. The map $z \mapsto 2z$ does the job. $\square$

We therefore get an isogeny of degree 4 of abelian surfaces over Y,

$$\phi \colon E^2 \to E \times_Y E^{(h)}, \quad (p, q) \mapsto (p, \tilde{h}(q)).$$

Note that if we had $h \in \mathrm{SL}_2(\mathbb{Z})$, then $\tilde{h}$ and ϕ would be isomorphisms.

By Deninger's method, we can relate $\mathrm{m}(P(x, y)P(z, t) - 16)$ to the motivic cohomology class $\xi = \{x, y, z, t\}$ on $E \times_Y E^{(h)}$. Note that the restriction of the polynomial $P(x, y)P(z, t) - 16$ to the torus vanishes only at the points $\pm(1, 1, 1, 1)$. Since the functions x, y, z, t are supported at 4-torsion points, the motivic class $\phi^*(\xi)$ should be related to Eisenstein symbols. We speculate that this can be converted into a 'modular' proof of the formula (10.8).

Chapter notes

Doran and Kerr [77] constructed motivic cohomology classes on Calabi–Yau varieties defined by families of toric hypersurfaces. These classes are, at least in some cases, related to Beilinson's Eisenstein symbols. Doran and Kerr also computed the images of these classes under the regulator map, and showed that they are related to the Mahler measures of the corresponding families of polynomials.

In the case of weight 2, formula (10.7) for $\int_0^\infty \eta(g_{a,b}, g_{c,d})$, together with the transformation formula for Siegel units, allows one to compute any integral of the form $\int_\alpha^\beta \eta(u, v)$, where u and v are arbitrary modular units and α and β are arbitrary cusps, in terms of L-values. In higher weight, Theorem 10.11 is not sufficient, as the symbols $\gamma(X^k\{0, \infty\})$ with $\gamma \in \mathrm{SL}_2(\mathbb{Z})$ do not span the space of weight $k + 2$ modular symbols when $k \geq 2$.

The method outlined in Section 10.5 could be used for other universal elliptic curves, at least for congruence subgroups Γ of $\mathrm{SL}_2(\mathbb{Z})$ such that the associated modular curve $Y(\Gamma)$ has genus 0. It would be interesting to find 'nice' (for example, tempered) equations for $\mathcal{E}(\Gamma)$ and deduce Mahler measure identities for the threefold $\mathcal{E}(\Gamma)^2$.

Additional exercises

Exercise 10.6 Let $N \geq 3$. Let $\mathcal{F}_N(\mathbb{C})$ be the set of isomorphism classes of pairs (E, α), where E is an elliptic curve over $\mathbb{C}$ and $\alpha: (\mathbb{Z}/N\mathbb{Z})^2 \xrightarrow{\cong} E[N]$ is an isomorphism of abelian groups. Show that $\mathcal{F}_N(\mathbb{C})$ is in bijection with $(\mathbb{Z}/N\mathbb{Z})^\times \times (\Gamma(N) \setminus \mathcal{H})$.

Hint Send $\tau \in \mathcal{H}$ to the pair (E_τ, α_τ) with $E_\tau = \mathbb{C}/(\mathbb{Z}\tau + \mathbb{Z})$ and $\alpha_\tau(m, n) = [(m\tau + n)/N]$. □

Exercise 10.7 Let $k \geq 1$. The *horospherical map* $\lambda^k: \mathbb{Q}[(\mathbb{Z}/N\mathbb{Z})^2] \to \mathbb{Q}[C_N]^{(k)}$ is defined by

$$\lambda^k(\phi)(g) = \sum_{v=(v_1,v_2)\in(\mathbb{Z}/N\mathbb{Z})^2} \phi(g^{-1}v)B_{k+2}\left(\left\{\frac{v_2}{N}\right\}\right)$$

where B_{k+2} is the Bernoulli polynomial, and $\{x\} = x - \lfloor x \rfloor$ is the fractional part of $x \in \mathbb{R}/\mathbb{Z}$.

(a) Check that λ^k is well defined.
(b) In the case $k = 0$, show that the same formula gives a map $\lambda^0: \mathbb{Q}[(\mathbb{Z}/N\mathbb{Z})^2 \setminus \{0\}] \to \mathbb{Q}[C_N]^{(0)}$.

(c) Show that $\mathrm{Eis}^k(u) = \mathrm{B}^k_{\mathcal{M}}(\lambda^k([u]))$.

(d) (*More difficult*) Show that the map λ^k is surjective.

Hint (d) Relate λ^k to Dirichlet L-values $L(\chi, k+2)$. $\qquad\qquad$ □

The aim of the following exercise is to show that for $N \geq 3$, the modular forms $e_{a,d}e_{b,-c} + e_{a,-d}e_{b,c}$ from (10.7) generate $M_2(\Gamma(N))$.

The *Eisenstein subspace* $\mathrm{Eis}_1(\Gamma(N)) \subset M_1(\Gamma(N))$ is the subspace generated by the Eisenstein series $e_{a,b}$ with $a, b \in \mathbb{Z}/N\mathbb{Z}$. We admit that the space of modular forms is the direct sum of the Eisenstein subspace and the space of cusp forms, that is, $M_1(\Gamma(N)) = \mathrm{Eis}_1(\Gamma(N)) \oplus S_1(\Gamma(N))$. Khuri-Makdisi [112] showed that, for $N \geq 3$, the linear map

$$\mathrm{Eis}_1(\Gamma(N)) \otimes_{\mathbb{C}} \mathrm{Eis}_1(\Gamma(N)) \to M_2(\Gamma(N)), \quad e_1 \otimes e_2 \mapsto e_1 e_2$$

is surjective.

Exercise 10.8 (a) Show that $e_{b,a} = e_{a,b}$ and $e_{-a,-b} = -e_{a,b}$.

(b) Let V be the subspace of $\mathrm{Eis}_1(\Gamma(N))$ generated by the $e_{a,b} + e_{a,-b}$ with $a, b \in \mathbb{Z}/N\mathbb{Z}$. Show that $e_{a,b} - e_{a,-b} \in V$ and deduce that $V = \mathrm{Eis}_1(\Gamma(N))$.

(c) Conclude that for $N \geq 3$, the modular forms $e_{a,d}e_{b,-c} + e_{a,-d}e_{b,c}$ with $a, b, c, d \in \mathbb{Z}/N\mathbb{Z}$ generate $M_2(\Gamma(N))$.

(d) Show that the modular forms $e_{a,d}e_{b,-c} + e_{a,-d}e_{b,c}$ with $ad - bc = 0$ generate $M_2(\Gamma_1(N))$.

Appendix

Motivic cohomology and regulators

This appendix originates from lectures by Jörg Wildeshaus given at the summer school on special values of L-functions (ÉNS Lyon, June 2014).

In Section A.1, we describe the construction of the category of geometrical motives following Voevodsky [213, Chapter 5, Section 2]. This enables one to define the motivic cohomology of smooth varieties. In Section A.2, we explain how the $\mathbb{A}^1$-homotopy theory of schemes by Morel and Voevodsky [143] gives a natural framework for constructing regulator maps. Finally, in Section A.3 we define Deligne–Beilinson cohomology following Burgos [55], using the homotopic method of Déglise and Mazzari [63].

The prerequisites for this appendix are rather different from the rest of the book. We will need some tools of homological algebra and the notion of derived and triangulated categories. The reader may consult for example [92].

A.1 Motivic cohomology

Motives can be thought of as an attempt to linearise the category of algebraic varieties. In essence, the motive of an algebraic variety X is an abstraction of the cohomology groups associated to X. Given a cohomology theory (like singular cohomology, étale cohomology, ...), there should be a functor from the category of motives to the relevant category (typically, vector spaces with additional structure) which sends the motive of X to the cohomology of X we started with. In this sense, motives can be seen as a kind of universal cohomology theory for algebraic varieties. Another fruitful idea comes from the observation that the cohomology groups of an algebraic variety sometimes break up into smaller pieces. One expects that such a decomposition should hold at the motivic level.

Various categories of motives have been defined. We explain here only the construction of the triangulated category of geometrical motives defined by Voevodsky. We refer the reader to [6, 7, 59, 136, 145, 180] for a wider panorama.

Let k be a field of characteristic 0. A variety over k is a separated scheme of finite type over k. We denote by Sm/k the category of smooth varieties over k.

Definition A.1 Let $X, Y \in \text{Sm}/k$ with X connected. The group $c(X, Y)$ of finite correspondences from X to Y is the free abelian group on the set of integral closed subschemes $W \subset X \times_k Y$ such that the projection $W \to X$ is finite and surjective. For any such W, we denote by $[W]$ its image in $c(X, Y)$.

For an arbitrary $X \in \text{Sm}/k$, we define $c(X, Y) = \bigoplus_i c(X_i, Y)$, where the X_i are the connected components of X.

For example, every morphism $f: X \to Y$ between smooth varieties defines a finite correspondence from X to Y. Namely, if X is connected, the graph $\Gamma_f \subset X \times_k Y$ satisfies the condition of Definition A.1, thus defining an element $[\Gamma_f]$ in $c(X, Y)$. In general, we define $[\Gamma_f]$ by taking the sum over the connected components of X.

Finite correspondences can be composed as follows:

$$c(X, Y) \otimes c(Y, Z) \to c(X, Z),$$

$$\alpha \otimes \beta \mapsto \beta \circ \alpha := (p_{XZ})_* \left(p_{XY}^* \alpha \bullet p_{YZ}^* \beta \right)$$

where $\bullet$ is the intersection product defined by Serre (see [145, Chapter 1]).

Exercise A.1 (a) Show that the cycles $p_{XY}^* \alpha$ and $p_{YZ}^* \beta$ intersect properly, and that the intersection is a finite correspondence from X to $Y \times_k Z$.

(b) Show that for any two morphisms $f: X \to Y$ and $g: Y \to Z$, we have $[\Gamma_g] \circ [\Gamma_f] = [\Gamma_{g \circ f}]$.

Definition A.2 The category $\text{SmCor}(k)$ is the category with the same objects as Sm/k, but whose morphisms are given by $\text{Hom}_{\text{SmCor}(k)}(X, Y) := c(X, Y)$, the composition being defined as above.

Note that we have a covariant functor $\text{Sm}/k \to \text{SmCor}(k)$ given by $X \mapsto X$, $f \mapsto [\Gamma_f]$. The category $\text{SmCor}(k)$ is additive: the direct sum of two objects $X, Y \in \text{Sm}/k$ is given by the disjoint union $X \sqcup Y$.

Ideally, we would like to define motives as objects of a suitable abelian category. Such a construction is not known yet, so we aim for a triangulated category instead. Since triangulated categories admit cohomological functors, this is enough for our purpose of defining the motivic cohomology associated to an algebraic variety (or more generally, to a motive). The triangulated structure then also gives the usual long exact sequences of cohomology groups.

We let $K^b(\text{SmCor}(k))$ be the homotopy category of bounded complexes in

SmCor(k). The objects are the bounded complexes, and two morphisms of complexes are identified whenever they are homotopic. This is a triangulated category.

Notation A.3 Given a complex $K^{\cdot}$ in an additive category, its translate $K[n]^{\cdot}$ is defined by $K[n]^i = K^{i+n}$.

Definition A.4 We denote by $T \subset K^b(\mathrm{SmCor}(k))$ the full triangulated subcategory, thick, generated by the following:

(a) for every $X \in \mathrm{Sm}/k$, the complex $X \times_k \mathbb{A}^1 \xrightarrow{\mathrm{pr}_X} X$;
(b) for every $X \in \mathrm{Sm}/k$ and every open covering $X = U \cup V$, the complex

$$U \cap V \xrightarrow{\alpha} U \oplus V \xrightarrow{\beta} X$$

where α (respectively, β) is given by the difference (respectively, sum) of the two natural inclusions.

The complexes of type (a) above will ensure that motivic cohomology is invariant under $\mathbb{A}^1$-homotopy — we may see the affine line $\mathbb{A}^1$ as an algebraic analogue of the unit interval $[0, 1]$ in topology. The complexes of type (b) will ensure that motivic cohomology satisfies the Mayer–Vietoris property for open subsets.

Definition A.5 (1) The category of *effective geometrical motives* $\mathrm{DM}_{\mathrm{gm}}^{\mathrm{eff}}(k)$ is the pseudo-abelian envelope of $K^b(\mathrm{SmCor}(k))/T$.
(2) For any smooth variety X/k, the motive $M_{\mathrm{gm}}(X) \in \mathrm{DM}_{\mathrm{gm}}^{\mathrm{eff}}(k)$ is the complex concentrated in degree 0 given by X.

In fact, we have a covariant functor $M_{\mathrm{gm}} \colon \mathrm{Sm}/k \to \mathrm{DM}_{\mathrm{gm}}^{\mathrm{eff}}(k)$ given by the composition of the functor $\mathrm{Sm}/k \to \mathrm{SmCor}(k)$ defined above, and the natural functor $\mathrm{SmCor}(k) \to \mathrm{DM}_{\mathrm{gm}}^{\mathrm{eff}}(k)$.

The pseudo-abelian envelope of a triangulated category is triangulated, by virtue of a theorem by Balmer and Schlichting [11]. It follows that $\mathrm{DM}_{\mathrm{gm}}^{\mathrm{eff}}(k)$ is a triangulated category. Moreover, it has a natural tensor structure, induced by $M_{\mathrm{gm}}(X) \otimes M_{\mathrm{gm}}(Y) := M_{\mathrm{gm}}(X \times_k Y)$. The unit object is the motive of a point $M_{\mathrm{gm}}(\mathrm{Spec}\, k)$, denoted also by $\mathbb{Z}(0)$ or simply $\mathbb{Z}$.

We may think of the triangulated category $\mathrm{DM}_{gm}(k, \Lambda)$ as the derived category of the hypothetical abelian category of motives (as we said, the latter category has not been constructed yet).

Let us describe the motive of the projective line. Let $x \in \mathbb{P}^1(k)$ be any rational point, and let $x_* \colon \mathbb{Z}(0) = M_{\mathrm{gm}}(\mathrm{Spec}\, k) \to M_{\mathrm{gm}}(\mathbb{P}^1_k)$ be the corresponding morphism. The structural morphism $f \colon \mathbb{P}^1_k \to \mathrm{Spec}\, k$ is a left inverse of x, hence $f_* : M_{\text{gm}}(\mathbb{P}^1_k) \to \mathbb{Z}(0)$ satisfies $f_* x_* = (f \circ x)_* = \text{id}$. Since the category $\text{DM}^{\text{eff}}_{\text{gm}}(k)$ is pseudo-abelian, we get a splitting

$$M_{\text{gm}}(\mathbb{P}^1_k) = \mathbb{Z}(0) \oplus \mathbb{L},$$

where $\mathbb{L}$ is called the Lefschetz motive.

Definition A.6 The Tate motive $\mathbb{Z}(1)$ is defined as $\mathbb{L}[-2]$.

Exercise A.2 (Motivic decomposition of $\mathbb{G}_m$) We denote by $\mathbb{G}_m = \mathbb{A}^1_k \setminus \{0\}$ the multiplicative group over k.

(a) Show that $M_{\text{gm}}(\mathbb{G}_m) \cong \mathbb{Z}(0) \oplus \mathbb{Z}(1)[1]$.
(b) Show that the involution $t \mapsto t^{-1}$ of $\mathbb{G}_m$ preserves this decomposition, and identify its effect on each summand.

Hint (a) Use the Mayer–Vietoris complex from Definition A.4 for the standard open covering of $\mathbb{P}^1_k$ by two copies of $\mathbb{A}^1_k$. $\qquad\qquad\square$

Definition A.7 The category of *geometrical motives* $\text{DM}_{\text{gm}}(k)$ is obtained from $\text{DM}^{\text{eff}}_{\text{gm}}(k)$ by $\otimes$-inverting $\mathbb{Z}(1)$ (equivalently, $\otimes$-inverting $\mathbb{L}$).

For any motive $M \in \text{DM}_{\text{gm}}(k)$ and any $n \in \mathbb{Z}$, we let $M(n) := M \otimes \mathbb{Z}(n)$.

It is instructive to compare $\text{DM}_{\text{gm}}(k)$ with other categories of motives. For example, consider the category $\text{Chow}(k)$ of Chow motives, which are associated to smooth projective varieties [145, Chapter 2]. Voevodsky has showed that there exists a fully faithful functor $\text{Chow}(k) \to \text{DM}_{\text{gm}}(k)$ which sends the Chow motive of X to $M(X)$ for every smooth projective variety X [213, Chapter 5, Proposition 2.1.4, Proposition 4.2.6].

For any commutative ring Λ, one can define Λ-linear versions of the above categories, replacing formal sums by formal Λ-linear combinations. The resulting triangulated categories are denoted by $\text{Chow}(k, \Lambda)$ and $\text{DM}_{\text{gm}}(k, \Lambda)$. As before, we have the Tate motives $\Lambda(n) \in \text{DM}_{\text{gm}}(k, \Lambda)$ for any $n \in \mathbb{Z}$. For any ring morphism $\Lambda \to \Lambda'$, we have a canonical functor $\text{DM}_{\text{gm}}(k, \Lambda) \to \text{DM}_{\text{gm}}(k, \Lambda')$ obtained by extending the coefficients.

We are now in a position to define the motivic cohomology of smooth varieties.

Definition A.8 Let X be a smooth variety over k. Let $i, j \in \mathbb{Z}$, and let Λ be any commutative ring. The motivic cohomology of X with coefficients in Λ is defined by

$$H^i_{\mathcal{M}}(X, \Lambda(j)) = \text{Hom}_{\text{DM}_{\text{gm}}(k,\Lambda)}(M_{\text{gm}}(X), \Lambda(j)[i]).$$

For any morphism $f\colon X \to Y$, the morphism $M_{\mathrm{gm}}(f)\colon M_{\mathrm{gm}}(X) \to M_{\mathrm{gm}}(Y)$ induces a Λ-linear map $f^*\colon H^i_{\mathcal{M}}(Y, \Lambda(j)) \to H^i_{\mathcal{M}}(X, \Lambda(j))$, called the pull-back by f.

Motivic cohomology groups are hard to compute in general, and it's even difficult to construct elements. Here, however, is a simple example. Let $X \in \mathrm{Sm}/k$, and let $f \in O(X)^\times$ be an invertible function. Viewing f as a morphism $X \to \mathbb{G}_m$, we may consider the composite map $M_{\mathrm{gm}}(X) \xrightarrow{M_{\mathrm{gm}}(f)} M_{\mathrm{gm}}(\mathbb{G}_m) \twoheadrightarrow \mathbb{Z}(1)[1]$. This defines an element $[f] \in H^1_{\mathcal{M}}(X, \mathbb{Z}(1))$. The map $f \mapsto [f]$ induces an isomorphism of groups $O(X)^\times \cong H^1_{\mathcal{M}}(X, \mathbb{Z}(1))$; see [136, Lecture 4].

By construction of the category $\mathrm{DM}_{\mathrm{gm}}(k)$, we have the following properties.

- For any smooth variety X/k and any $i, j \in \mathbb{Z}$, we have a natural isomorphism
$$H^i_{\mathcal{M}}(X, \Lambda(j)) \cong H^i_{\mathcal{M}}(X \times_k \mathbb{A}^1, \Lambda(j)).$$

- For any open covering $X = U \cup V$, we have a long exact sequence
$$\cdots \to H^i_{\mathcal{M}}(X, \Lambda(j)) \to H^i_{\mathcal{M}}(U, \Lambda(j)) \oplus H^i_{\mathcal{M}}(V, \Lambda(j)) \to H^i_{\mathcal{M}}(U \cap V, \Lambda(j))$$
$$\to H^{i+1}_{\mathcal{M}}(X, \Lambda(j)) \cdots.$$

Exercise A.3 Using Exercise A.2, show that for any smooth variety X over k, we have
$$H^i_{\mathcal{M}}(X \times_k \mathbb{G}_m, \Lambda(j)) \cong H^i_{\mathcal{M}}(X, \Lambda(j)) \oplus H^{i-1}_{\mathcal{M}}(X, \Lambda(j-1)).$$

Finally, let us discuss products. Let $\alpha \in H^i_{\mathcal{M}}(X, \Lambda(j))$ and $\beta \in H^{i'}_{\mathcal{M}}(X, \Lambda(j'))$ be two motivic classes on X. The cup-product $\alpha \cup \beta \in H^{i+i'}_{\mathcal{M}}(X, \Lambda(j + j'))$ is defined as follows. We view our classes as morphisms $\alpha\colon M_{\mathrm{gm}}(X) \to \Lambda(j)[i]$ and $\beta\colon M_{\mathrm{gm}}(X) \to \Lambda(j')[i']$. Then $\alpha \cup \beta$ is defined as the composition
$$M_{\mathrm{gm}}(X) \xrightarrow{\Delta_*} M_{\mathrm{gm}}(X \times_k X) = M_{\mathrm{gm}}(X) \otimes M_{\mathrm{gm}}(X) \xrightarrow{\alpha \otimes \beta} \Lambda(j + j')[i + i'],$$
where $\Delta\colon X \to X \times_k X$ is the diagonal embedding.

Exercise A.4 (a) Let τ be the transposition on $\mathbb{P}^1_k \times \mathbb{P}^1_k$, and let $M_{\mathrm{gm}}(\tau)$ be the induced morphism on $M_{\mathrm{gm}}(\mathbb{P}^1_k \times \mathbb{P}^1_k) = M_{\mathrm{gm}}(\mathbb{P}^1_k) \otimes M_{\mathrm{gm}}(\mathbb{P}^1_k)$. Show that $M_{\mathrm{gm}}(\tau)$ acts trivially on $\mathbb{L} \otimes \mathbb{L}$.

(b) Deduce that the cup-product is graded commutative: using the notation above, we have $\beta \cup \alpha = (-1)^{ii'} \alpha \cup \beta$.

A.2 Regulator maps

As before, k is a field of characteristic 0.

We now explain how to construct regulator maps from motivic cohomology to other cohomology theories. The main tool is $\mathbb{A}^1$-homotopy theory for schemes [143]. The idea is to embed the category of geometrical motives in the so-called $\mathbb{A}^1$-derived category, where the objects have an 'explicit description' and can be constructed by hand.

More precisely, assume that Λ is a $\mathbb{Q}$-algebra. There is a fully faithful functor from $\mathrm{DM}_{\mathrm{gm}}(k, \Lambda)$ to the category of *Morel motives* $\mathrm{D}_{\mathbb{A}^1}(k, \Lambda)_+ \subset \mathrm{D}_{\mathbb{A}^1}(k, \Lambda)$. Assume that we are given a Λ-algebra object $\mathbb{E}$ in the category of Morel motives. We then define the $\mathbb{E}$-cohomology of a smooth variety X/k by

$$H^i_{\mathbb{E}}(X, \Lambda(j)) = \mathrm{Hom}_{\mathrm{D}_{\mathbb{A}^1}(k,\Lambda)}(M_{\mathrm{gm}}(X), \mathbb{E}(j)[i]).$$

The $\mathbb{E}$-cohomology is functorial and thanks to the ring structure on $\mathbb{E}$, it also admits products. Moreover, the structural morphism $\Lambda(0) \to \mathbb{E}$ induces a Λ-linear map

$$\mathrm{reg}_{\mathbb{E}} \colon H^i_{\mathcal{M}}(X, \Lambda(j)) \to H^i_{\mathbb{E}}(X, \Lambda(j))$$

called the $\mathbb{E}$-regulator. It is functorial and compatible with products. It turns out that the Deligne–Beilinson cohomology of smooth varieties over $\mathbb{R}$ with real coefficients can be represented by an $\mathbb{R}$-algebra object $\mathbb{DB} \in \mathrm{D}_{\mathbb{A}^1}(\mathbb{R}, \mathbb{R})_+$ (see Section A.3). This immediately gives the existence of the Beilinson regulator map and its compatibility with the various operations in cohomology.

Let us fix some notation. As before, Sm/k is the category of smooth varieties over k, and Λ is a commutative ring. Let $\mathrm{PSh}(k, \Lambda)$ be the abelian category of presheaves of Λ-modules on Sm/k. Let $\mathrm{C}(\mathrm{PSh}(k, \Lambda))$ be the category of (unbounded) complexes in $\mathrm{PSh}(k, \Lambda)$, and let $\mathrm{D}(\mathrm{PSh}(k, \Lambda))$ be the associated derived category. The effective $\mathbb{A}^1$-derived category $\mathrm{D}^{\mathrm{eff}}_{\mathbb{A}^1}(k, \Lambda)$ is a certain full triangulated subcategory of $\mathrm{D}(\mathrm{PSh}(k, \Lambda))$. To define it, we need to introduce the Nisnevich topology (see [143, 3.1], [136, Lecture 12]).

Definition A.9 Let X be a smooth k-variety. A *Nisnevich covering of X* is a finite family of étale morphisms $(f_i \colon U_i \to X)_{i \in I}$ in Sm/k having the following lifting property: for every $x \in X$, there exists an $i \in I$ and $u \in U_i$ such that $f_i(u) = x$ and the extension of residue fields $k(x) \to k(u)$ is an isomorphism.

The Nisnevich coverings define a Grothendieck topology on the category Sm/k. This topology is finer than the Zariski topology, but coarser than the étale topology, as the following examples illustrate.

Example A.10 (1) Any Zariski open covering $X = \bigcup_{i \in I} U_i$ is also a Nisnevich covering.

(2) By definition, every Nisnevich covering is an étale covering. The converse is not true: consider the étale covering $f \colon \mathbb{A}^1_k \setminus \{0\} \to \mathbb{A}^1_k \setminus \{0\}$ given by

$t \mapsto t^2$. The fibre above the generic point is given by the field extension $k(u) \to k(u^{1/2})$, which has degree 2. Hence f is not a Nisnevich covering. We can get a Nisnevich covering by adding to f the open immersion $j \colon \mathbb{A}^1_k \setminus \{0, 1\} \to \mathbb{A}^1_k \setminus \{0\}$.

A useful property of the Nisnevich topology is that, in contrast with the étale topology, it is generated by simple coverings.

Definition A.11 A *distinguished square* is a commutative diagram in Sm/k,

$$
\begin{array}{ccc}
W & \longhookrightarrow & V \\
\downarrow & & \downarrow f \\
U & \underset{j}{\longhookrightarrow} & X,
\end{array}
\tag{A.1}
$$

where j is an open embedding, f is étale, $W = f^{-1}(U)$ and $V \setminus W \to X \setminus U$ is an isomorphism of reduced schemes.

With the notation as in (A.1), the family $\{U \to X, V \to X\}$ is a Nisnevich covering of X (note that this generalises the usual Zariski coverings $X = U \cup V$). Moreover, the Nisnevich topology is generated by the distinguished squares [143, Proposition 1.4].

Definition A.12 Let $K \in C(\mathrm{PSh}(k, \Lambda))$. We say that

(a) K is Nis-local if $K(\emptyset)$ is acyclic, and every distinguished square (A.1) gives rise canonically to a long exact sequence

$$\cdots H^{n-1}(K(W)) \to H^n(K(X)) \to H^n(K(U)) \oplus H^n(K(V)) \to H^n(K(W)) \cdots$$

(see [58, 1.1.9] for more details on this property) and

(b) K is $\mathbb{A}^1$-local if for every $X \in \mathrm{Sm}/k$, the natural map of complexes $K(X) \to K(X \times_k \mathbb{A}^1)$ is a quasi-isomorphism.

Working with Nis-local and $\mathbb{A}^1$-local objects will allow us to define cohomology theories with good descent properties, as properties (a) and (b) above indicate. This leads to the following definition.

Definition A.13 The *effective $\mathbb{A}^1$-derived category* over k with coefficients in Λ, denoted by $D^{\mathrm{eff}}_{\mathbb{A}^1}(k, \Lambda)$, is the full subcategory of Nis-local, $\mathbb{A}^1$-local objects in $D(\mathrm{PSh}(k, \Lambda))$.

Let us give some examples with explicit presheaves.

Definition A.14 Let $X \in \mathrm{Sm}/k$. The presheaf $\Lambda(X)$ represented by X is defined by $Y \mapsto \Lambda \mathrm{Hom}_k(Y, X)$, where ΛB denotes the free Λ-module generated by a set B.

For example, the presheaf $\Lambda(\mathrm{Spec}\,k)$ represented by a point $\mathrm{Spec}\,k$ is the constant presheaf with value Λ. We denote it simply by Λ.

Exercise A.5 Show that $\Lambda(X \times Y) \cong \Lambda(X) \otimes \Lambda(Y)$ for any smooth varieties X, Y over k.

Exercise A.6 (a) Show that the presheaf $\Lambda(\mathbb{A}^1_k)$ (seen as a complex concentrated in degree 0) is not $\mathbb{A}^1$-local.

(b) Show that the presheaf $\Lambda(\mathbb{G}_m)$ is $\mathbb{A}^1$-local, but not Nis-local.

(c) Same question with $\mathbb{G}_m$ replaced by a smooth curve of genus ≥ 1.

As these exercises show, many natural presheaves are neither Nis-local nor $\mathbb{A}^1$-local. The following theorem gives a way to construct objects in $D^{\mathrm{eff}}_{\mathbb{A}^1}(k, \Lambda)$.

Theorem A.15 *The inclusion functor* $D^{\mathrm{eff}}_{\mathbb{A}^1}(k, \Lambda) \to D(\mathrm{PSh}(k, \Lambda))$ *admits a left adjoint* $a_{\mathbb{A}^1} : D(\mathrm{PSh}(k, \Lambda)) \twoheadrightarrow D^{\mathrm{eff}}_{\mathbb{A}^1}(k, \Lambda)$.

The functor $a_{\mathbb{A}^1}$ induces a map $C(\mathrm{PSh}(k, \Lambda)) \twoheadrightarrow D^{\mathrm{eff}}_{\mathbb{A}^1}(k, \Lambda)$, which one may view as a kind of localisation. In fact, this can be made precise: $D^{\mathrm{eff}}_{\mathbb{A}^1}(k, \Lambda)$ identifies with the homotopy category of $C(\mathrm{PSh}(k, \Lambda))$ with respect to a certain explicit model structure [58, Proposition 1.1.15].

From now on, we will view objects of $C(\mathrm{PSh}(k, \Lambda))$, like $\Lambda(X)$, as objects of $D^{\mathrm{eff}}_{\mathbb{A}^1}(k, \Lambda)$ via $a_{\mathbb{A}^1}$.

We now define the analogue of the Tate motive from Section A.1. Recall that by Exercise A.2, the Tate motive $\Lambda(1)$ identifies with the cokernel of the morphism $1_* : M_{\mathrm{gm}}(\mathrm{Spec}\,k) \to M_{\mathrm{gm}}(\mathbb{G}_m)$, seen as a complex concentrated in degree 1. We make a similar definition here.

Definition A.16 The Tate object is $\Lambda(1) := a_{\mathbb{A}^1}(\mathrm{coker}(\Lambda \to \Lambda(\mathbb{G}_m))[-1])$.

The category $D^{\mathrm{eff}}_{\mathbb{A}^1}(k, \Lambda)$ is endowed with a tensor structure [58, Proposition 1.2.2]. For any object K and any integer $n \geq 0$, we define $K(n) := K \otimes \Lambda(1)^{\otimes n}$.

The Tate object $\Lambda(1)$ is not $\otimes$-invertible (in other words, the functor $K \mapsto K(1)$ is not an equivalence of the category $D^{\mathrm{eff}}_{\mathbb{A}^1}(k, \Lambda)$). It is possible to modify the construction of $D^{\mathrm{eff}}_{\mathbb{A}^1}(k, \Lambda)$ to define the $\mathbb{A}^1$-derived category $D_{\mathbb{A}^1}(k, \Lambda)$ in which $\Lambda(1)$ will be $\otimes$-invertible. There is a pair of adjoint functors

$$\Sigma^\infty : D^{\mathrm{eff}}_{\mathbb{A}^1}(k, \Lambda) \underset{\longleftarrow}{\overset{\longrightarrow}{\rule{0pt}{0pt}}} D_{\mathbb{A}^1}(k, \Lambda) : \Omega^\infty.$$

We refer to [58, 1.4] and [59, Section 5.3] for the actual construction, which is quite involved and requires symmetric Tate spectra, an idea borrowed from algebraic topology.

We finally come to the construction of regulator maps. As we explained at the beginning of this section, there are two main ingredients. The first is the following embedding theorem due to Morel.

Theorem A.17 *Assume that Λ is a $\mathbb{Q}$-algebra. Then there exists a canonical fully faithful functor $\mathrm{DM}_{\mathrm{gm}}(k, \Lambda) \hookrightarrow D_{\mathbb{A}^1}(k, \Lambda)_+$ mapping $M_{\mathrm{gm}}(X)$ to $\Sigma^\infty \Lambda(X)$ for every smooth variety X over k.*

Given any object $\mathbb{E}$ of $D_{\mathbb{A}^1}(k, \Lambda)$, the $\mathbb{E}$-cohomology of a smooth variety X over k is defined by

$$H_{\mathbb{E}}^i(X, \Lambda(j)) := \mathrm{Hom}_{D_{\mathbb{A}^1}(k,\Lambda)}(\Sigma^\infty \Lambda(X), \Lambda(j)[i]).$$

The second ingredient in the regulator map is the construction of a ring object in the $\mathbb{A}^1$-derived category representing the cohomology theory we are interested in. We give here a method due to Déglise and Mazzari [63].

Theorem A.18 ([63, 1.4.10]) *Let $(E_j)_{j \geq 0}$ be a sequence of Nis-local, $\mathbb{A}^1$-local objects in $C(\mathrm{PSh}(k, \Lambda))$ equipped with*

- *$\eta \colon \Lambda \to E_0$,*
- *$\mu_{i,j} \colon E_i \otimes E_j \to E_{i+j}$ for every $i, j \geq 0$,*

such that the following diagrams 'unit' and 'associativity', respectively,

$$
\begin{array}{ccc}
E_j \xrightarrow{\mathrm{id} \otimes \eta} E_j \otimes E_0 & \qquad E_i \otimes E_j \otimes E_k \longrightarrow E_i \otimes E_{j+k} \\
\searrow \quad \Big\downarrow{\mu_{j,0}} & \qquad \Big\downarrow \qquad\qquad\qquad \Big\downarrow \\
\quad E_j, & \qquad E_{i+j} \otimes E_k \longrightarrow E_{i+j+k},
\end{array}
$$

commute in $D(\mathrm{PSh}(k, \Lambda))$, and the diagram 'commutativity' commutes in $C(\mathrm{PSh}(k, \Lambda))$.

Suppose also that we are given $c \in H^1(E_1(\mathbb{G}_m))$ satisfying the following stability condition: for every $X \in \mathrm{Sm}/k$, $i \in \mathbb{Z}$ and $j \geq 0$, we have an isomorphism

$$H^i(E_j(X)) \oplus H^{i+1}(E_{j+1}(X)) \xrightarrow[\cong]{\alpha_1 \oplus \alpha_0} H^{i+1}(E_{j+1}(X \times_k \mathbb{G}_m)), \tag{A.2}$$

where the map α_1 is defined by taking the product with c, and the map α_0 is induced by the projection $X \times_k \mathbb{G}_m \to X$.

Then there exists a Λ-algebra $\mathbb{E}$ in $D_{\mathbb{A}^1}(k, \Lambda)$ such that for every $i \in \mathbb{Z}$ and every integer $j \geq 0$, we have

$$H_{\mathbb{E}}^i(X, \Lambda(j)) \cong H^i(E_j(X)).$$

Note that the $\mathbb{E}$-cohomology $H_{\mathbb{E}}^i(X, \Lambda(j))$ is given by the cohomology of a single complex of Λ-modules $E_j(X)$. Note also that the stability condition (A.2) is similar to the decomposition appearing in Exercise A.3, which itself comes from the motivic decomposition of $\mathbb{G}_m$.

With the notation as in Theorem A.18, let us assume in addition that $u^*(c) = -c \bmod H^1(E_1(\operatorname{Spec} k))$, where u is the involution of $\mathbb{G}_m$ defined by $t \mapsto t^{-1}$. Then $\mathbb{E}$ is a Morel motive, in other words $\mathbb{E} \in D_{\mathbb{A}^1}(k, \Lambda)_+$. In this case, the $\mathbb{E}$-cohomology is graded commutative. We also say that $\mathbb{E}$ is a motivic ring spectrum. As explained at the beginning of this section, the $\mathbb{E}$-regulator is then simply defined using the unit map $\eta \colon \Lambda \to \mathbb{E}$. More precisely, for any smooth variety X over k, the regulator map from the motivic cohomology of X to the $\mathbb{E}$-cohomology of X is defined as

$$H^i_{\mathcal{M}}(X, \Lambda(j)) = \operatorname{Hom}_{\mathrm{DM}_{\mathrm{gm}}(k,\Lambda)}(M_{\mathrm{gm}}(X), \Lambda(j)[i])$$
$$\to \operatorname{Hom}_{D_{\mathbb{A}^1}(k,\Lambda)}(\Sigma^\infty \Lambda(X), \mathbb{E}(j)[i])$$
$$= H^i_{\mathbb{E}}(X, \Lambda(j)).$$

We refer to [63] for a list of properties of the $\mathbb{E}$-cohomology and $\mathbb{E}$-regulator, when $\mathbb{E}$ is a motivic ring spectrum. In particular, we also have a theory of Borel–Moore $\mathbb{E}$-homology, and together with $\mathbb{E}$-cohomology they form a Poincaré duality theory with supports in the sense of [29, 106]. Let us also mention that it is possible to extend the definition of $\mathbb{E}$-cohomology to singular varieties, and that such an extension is unique [63, 2.2.11].

In the next section we explain the construction of the E_j for the Deligne–Beilinson cohomology with real coefficients. In this case the complexes $E_j(X)$ are very explicit and quite amenable to computations.

A.3 Deligne–Beilinson cohomology

In this section we define the Deligne–Beilinson cohomology groups associated to a non-singular complex algebraic variety, by constructing a spectrum in the sense of Section A.2. The complexes appearing here were introduced by Burgos [55]. The Deligne–Beilinson spectrum was constructed by Holmstrom and Scholbach [102], but here we follow the recipe of Theorem A.18.

Let $X \in \mathrm{Sm}/\mathbb{C}$ be a non-singular complex algebraic variety. In order to define its Deligne–Beilinson cohomology, we need the existence of a 'good' compactification, which means an open embedding $j \colon X \hookrightarrow \overline{X}$ where $\overline{X}$ is a smooth proper variety and $D = \overline{X} \setminus j(X)$ is a normal crossings divisor in $\overline{X}$.

This can be achieved using Hironaka's embedded resolution of singularities, in the following way. By Nagata's compactification theorem, there exists an open embedding of X into a proper variety $\tilde{X}$. Let $Z = \tilde{X} \setminus X$ be the complement, seen as a reduced closed subscheme of $\tilde{X}$. Applying Hironaka's theorem to the closed immersion $i \colon Z \to \tilde{X}$, we get a smooth proper variety $\overline{X}$ together with a proper birational map $p\colon \overline{X} \to \tilde{X}$ which is an isomorphism above X and such that $p^{-1}(Z)$ is a normal crossings divisor in $\overline{X}$. The composite map $j\colon X \cong p^{-1}(X) \hookrightarrow \overline{X}$ gives the desired compactification.

In Definition 8.1, we introduced the differential forms with logarithmic singularities along D. We slightly change notation since we want to keep track of the divisor D.

Notation A.19 For $\Lambda \in \{\mathbb{R}, \mathbb{C}\}$, we denote by $E^n_{X,\Lambda}(\log D)$ the space of Λ-valued C^∞ differential n-forms on X with logarithmic singularities along D.

The definition above being local, we actually have a sheaf on $\overline{X}$ for the usual topology, whose global sections are given by $E^n_{X,\Lambda}(\log D)$. The collection of these sheaves provides a fine resolution of the sheaf $j_*\Lambda$ on $\overline{X}$. Hence their global sections compute the singular (or de Rham) cohomology of X:

$$H^n(E^*_{X,\Lambda}(\log D)) \cong H^n_{\mathrm{sing}}(X, \Lambda). \tag{A.3}$$

The complex structure on X gives rise to a decomposition

$$E^n_{X,\mathbb{C}}(\log D) = \bigoplus_{p+q=n} E^{p,q}_{X,\mathbb{C}}(\log D), \tag{A.4}$$

where $E^{p,q}$ denotes the subspace of forms of type (p,q). We can then define a filtration

$$F^p E^n_{X,\mathbb{C}}(\log D) = \bigoplus_{\substack{p'+q'=n \\ p' \geq p}} E^{p',q'}_{X,\mathbb{C}}(\log D).$$

This induces, via the isomorphism (A.3), the Hodge filtration on $H^n(X, \mathbb{C})$ defined by Deligne [65]. The weight filtration can also be recovered, by assigning weight 0 to the regular forms on $\overline{X}$, and weight 1 to the local sections $\log|z_i|$, $\frac{dz_i}{z_i}$ and $\frac{d\overline{z_i}}{\overline{z_i}}$.

Since we want functoriality with respect to morphisms of smooth varieties, we need to work with spaces that are independent of the compactification. Given two good compactifications $\overline{X_1}$ and $\overline{X_2}$ of X, there exists a common refinement $\overline{X}$, fitting in a commutative diagram

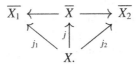

Such an $\overline{X}$ is obtained by resolving the singularities of the closure of the image of the diagonal embedding $X \to \overline{X_1} \times \overline{X_2}$. Thus, the category formed by

the good compactifications of X is directed, and we can make the following definition.

Definition A.20 Let $X \in \mathrm{Sm}/\mathbb{C}$ and $\Lambda \in \{\mathbb{R}, \mathbb{C}\}$. The complex of differential forms on X with logarithmic singularities at infinity is defined by

$$E_{\log,\Lambda}^*(X) = \varinjlim_{j:\, X \hookrightarrow \overline{X}} E_{\overline{X},\Lambda}^*(\log D),$$

where the direct limit is over the good compactifications $j: X \hookrightarrow \overline{X}$.

Taking the direct limit in (A.4), we get a decomposition

$$E_{\log,\mathbb{C}}^n(X) = \bigoplus_{p+q=n} E_{\log,\mathbb{C}}^{p,q}(X),$$

and thus a filtration on $E_{\log,\mathbb{C}}^n(X)$. The exterior derivative $\mathrm{d}: E^n \to E^{n+1}$ decomposes as $\mathrm{d} = \partial + \overline{\partial}$ with $\partial: E^{p,q} \to E^{p+1,q}$ and $\overline{\partial}: E^{p,q} \to E^{p,q+1}$.

By [54, Theorem 1.2], the canonical maps $E_{\overline{X},\mathbb{C}}^*(\log D) \to E_{\log,\mathbb{C}}^*(X)$ are real filtered quasi-isomorphisms. Hence the Deligne–Beilinson cohomology groups below will not depend on the chosen compactification of X.

In order to apply Theorem A.18, we need the complexes E_j. We define them as in Definition 8.3 but using the canonical spaces $E_{\log,\mathbb{C}}^n$ defined above instead.

Let us show that E_j is a complex of presheaves on $\mathrm{Sm}/\mathbb{C}$, in other words that $E_j(X)$ is contravariant in X. Let $f: X \to Y$ be a morphism between smooth complex varieties. Let us show that X and Y admit good compactifications $\overline{X}$ and $\overline{Y}$ such that f extends to a morphism $\overline{f}: \overline{X} \to \overline{Y}$. Note that any such extension will satisfy $\overline{f}^{-1}(\overline{Y} \setminus Y) \subset \overline{X} \setminus X$. The proof goes as follows. Let $\overline{Y}$ be any good compactification of Y. By Nagata's compactification theorem, the morphism $X \to \overline{Y}$ can be written as the composition of an open embedding $X \hookrightarrow X'$ and a proper morphism $X' \to \overline{Y}$. Applying embedded resolution of singularities to the closed immersion $X' \setminus X \to X'$ and reasoning as before, we get a good compactification $X \hookrightarrow \overline{X}$ together with a morphism $\overline{f}: \overline{X} \to \overline{Y}$ extending f. Now that we have $\overline{f}$, the crucial point is to see that the pull-back by f of a differential form on Y with logarithmic singularities along $D_Y = \overline{Y} \setminus Y$ has logarithmic singularities along $D_X = \overline{X} \setminus X$. This can be shown by taking local analytic coordinates adapted to D_X (respectively, D_Y) and using the analytic Nullstellensatz [104, Proposition 1.1.29] and the unique factorisation property for germs of analytic functions [104, Proposition 1.1.15].

We also need to construct the unit η, the multiplication $\mu_{j,k}$ and the class c. The unit η is given by the obvious map $\mathbb{R} \to E_0(X)^0$. The class c is constructed in the following exercise.

Exercise A.7 Let $X = \mathbb{C}^\times$ be the multiplicative group over $\mathbb{C}$.

(a) Show that $E_1(X)$ is the complex $E^0_{\log,\mathbb{R}}(\mathbb{C}^\times) \xrightarrow{-2\partial\bar\partial} (2\pi i)E^2_{\log,\mathbb{R}}(\mathbb{C}^\times)$, placed in degrees 1 and 2.

(b) We choose the compactification $\overline{X} = \mathbb{P}^1(\mathbb{C})$, so that $D = \{0, \infty\}$. Show that the $\mathbb{R}$-algebra $E^0_{X,\mathbb{R}}(\log D)$ is equal to $C^\infty(\overline{X})[\log|z|]$, where $C^\infty(\overline{X})$ denotes the real-valued C^∞ functions on $\overline{X}$.

(c) Show that the class $c := \log|z|$ defines an element of $H^1(E_1(X))$.

(d) Show that the group $H^1(E_1(X))$ consists of those functions of the form $z \mapsto a + b\log|z|$ with $a, b \in \mathbb{R}$.

(e) Describe similarly the groups $E^1_{X,\mathbb{R}}(\log D)$ and $E^2_{X,\mathbb{R}}(\log D)$, and determine the group $H^2(E_1(X))$.

Note that the class c constructed above is anti-invariant under the involution $z \mapsto z^{-1}$ of $\mathbb{G}_m$. Hence, the Deligne–Beilinson spectrum will be a Morel motive. Finally, we define the multiplication map $\mu_{j,k}: E_j \otimes E_k \to E_{j+k}$.

Notation A.21 For a differential form $\alpha = \sum_{p+q=n-1} \alpha^{p,q} \in E^{n-1}_{\log,\mathbb{C}}$, write

$$\pi_j(\alpha) = \frac{1}{2}(\alpha + (-1)^j\bar\alpha),$$

$$\mathrm{pr}_j(\alpha) = \sum_{p,q<j} \alpha^{p,q} \quad \text{and} \quad F^{j,j}(\alpha) = \sum_{p,q\geq j} \alpha^{p,q},$$

$$r_j(\alpha) = \begin{cases} \partial(\alpha^{j-1,n-j}) - \bar\partial(\alpha^{n-j,j-1}) & \text{if } n \leq 2j-1, \\ \alpha & \text{if } n \geq 2j. \end{cases}$$

Definition A.22 Fix integers $j, k \geq 0$. Write $\ell = j + k$. For $\alpha \in E_j(X)^n$ and $\beta \in E_k(X)^m$, we define

$$\mu_{j,k}(\alpha \otimes \beta) = \begin{cases} (-1)^n r_j(\alpha) \wedge \beta + \alpha \wedge r_k(\beta) \\ \qquad \text{for } n < 2j, \ m < 2k, \\[4pt] \mathrm{pr}_\ell(\alpha \wedge \beta) \\ \qquad \text{for } n < 2j, \ m \geq 2k, \ n+m < 2\ell, \\[4pt] F^{\ell,\ell}(r_j(\alpha) \wedge \beta) + 2\pi_\ell\partial((\alpha \wedge \beta)^{\ell-1,n+m-\ell}) \\ \qquad \text{for } n < 2j, \ m \geq 2k, \ n+m \geq 2\ell, \\[4pt] \alpha \wedge \beta \\ \qquad \text{for } n \geq 2j, \ m \geq 2k, \\[4pt] + \text{ graded commutative prolongation} \\ \qquad \text{for } n \geq 2j, \ m < 2k. \end{cases}$$

The diagrams 'unit' and 'commutativity' from Theorem A.18 commute. By [55, 3.3], the diagram 'associativity' also commutes up to a natural homotopy. In particular, it commutes in $D(\mathrm{PSh}(\mathbb{C}, \mathbb{R}))$.

It remains to show that the E_j are Nis-local, $\mathbb{A}^1$-local, and that c satisfies the stability condition. To that end, we relate Deligne–Beilinson cohomology to singular and de Rham cohomology.

Exercise A.8 Let $j \geq 0$ be an integer.

(a) Check that the map r_j (see Notation A.21) defines a morphism of complexes $r_j \colon E_j \to (2\pi i)^j E^*_{\log, \mathbb{R}}$ in $C(\mathrm{PSh}(\mathbb{C}, \mathbb{R}))$.

(b) Let $\mathrm{can}_j \colon (2\pi i)^j E^*_{\log, \mathbb{R}} \to E^*_{\log, \mathbb{C}} / F^j E^*_{\log, \mathbb{C}}$ be the canonical morphism. Show that $\mathrm{can}_j \circ r_j$ is null homotopic, and hence is zero in $D(\mathrm{PSh}(\mathbb{C}, \mathbb{R}))$.

By [55, 2.6.2], the morphisms r_j and can_j actually fit into a distinguished triangle in $D(\mathrm{PSh}(\mathbb{C}, \mathbb{R}))$. We thus get a functorial long exact sequence

$$\cdots \to H^{n-1}_{\mathrm{dR}}(X, \mathbb{C}) / F^j H^{n-1}_{\mathrm{dR}}(X, \mathbb{C}) \to H^n(E_j(X)) \xrightarrow{r_j} H^n_{\mathrm{sing}}(X, (2\pi i)^j \mathbb{R}) \qquad \text{(A.5)}$$

$$\to H^n_{\mathrm{dR}}(X, \mathbb{C}) / F^j H^n_{\mathrm{dR}}(X, \mathbb{C}) \to \cdots .$$

Note that $H^{n-1}_{\mathrm{dR}}(X, \mathbb{C})$ is isomorphic to the algebraic de Rham cohomology of X, by Grothendieck's theorem [96]. Since the cohomology functors H^*_{dR}, $F^j H^*_{\mathrm{dR}}$ and H^*_{sing} satisfy Nis-descent (and even étale descent) and are $\mathbb{A}^1$-homotopy invariant, it follows from the sequence above that the E_j are Nis-local and $\mathbb{A}^1$-local.

To prove the stability of c, we use the following exercise.

Exercise A.9 (a) Using the recipe of Theorem A.18, prove that there exist spectra $\mathbb{B}$ and $\mathbb{FDR}$ (the letters standing respectively for Betti and filtered de Rham) such that for every $X \in \mathrm{Sm}/\mathbb{C}$, we have

$$H^n_{\mathbb{B}}(X, \mathbb{R}(j)) = H^n_{\mathrm{sing}}(X, (2\pi i)^j \mathbb{R}),$$

$$H^n_{\mathbb{FDR}}(X, \mathbb{R}(j)) = F^j H^n_{\mathrm{dR}}(X).$$

(b) Show that $\mathbb{B}$ and $\mathbb{FDR}$ are Morel motives.

(c) Show that for $X = \mathbb{G}_m$, the map r_1 sends c to $c_{\mathbb{B}}$.

Since stability is satisfied for $c_{\mathbb{B}}$ and c_{FDR}, and since the long exact sequence (A.5) is compatible with taking the product with c, we obtain the stability of c.

Altogether, the assumptions of Theorem A.18 are satisfied for the E_j, and we get the Deligne–Beilinson spectrum $\mathbb{DB} = (E_j)_{j \geq 0} \in D_{\mathbb{A}^1}(\mathbb{C}, \mathbb{R})_+$. For every $X \in \mathrm{Sm}/\mathbb{C}$, $i, j \in \mathbb{Z}$ with $j \geq 0$, we have

$$H^i_{\mathbb{DB}}(X, \mathbb{R}(j)) = H^i(E_j(X)).$$

The sequence (A.5) shows that the Deligne–Beilinson cohomology groups are finite-dimensional real vector spaces.

The construction can be modified to give the Deligne–Beilinson spectrum for varieties defined over $\mathbb{R}$, by considering the complex points and taking the invariants under the de Rham conjugation, as in Section 8.1. This gives complexes $E_j \in C(\mathrm{PSh}(\mathbb{R},\mathbb{R}))$, to which we may apply Theorem A.18, providing a motivic ring spectrum $\mathbb{DB} \in D_{\mathbb{A}^1}(\mathbb{R},\mathbb{R})_+$. Similarly to (A.5), we have long exact sequences

$$\cdots \to \left(H^{n-1}_{\mathrm{dR}}(X(\mathbb{C}),\mathbb{C})/F^j \right)^{F_{\mathrm{dR}}=1} \to H^n_{\mathrm{DB}}(X_{\mathbb{R}},\mathbb{R}(j))$$
$$\xrightarrow{r_j} H^n_{\mathrm{sing}}(X(\mathbb{C}),(2\pi i)^j\mathbb{R})^{F^*_\infty=(-1)^j} \to \cdots .$$

It is possible to define Deligne–Beilinson cohomology with Λ-coefficients, where Λ is any subring of $\mathbb{R}$ [55]. The resulting cohomology groups combine two pieces of information: the Λ-structure and the Hodge structure on de Rham cohomology. In fact, the Deligne–Beilinson cohomology groups with Λ-coefficients are isomorphic to extension groups in the category of mixed Hodge Λ-structures [16].

We finish with an exercise giving an example where the regulator map can be computed explicitly.

Exercise A.10 Let $X \in \mathrm{Sm}/\mathbb{C}$, and let $f \in O(X)^\times$ be an invertible function. Show that the regulator map $\mathrm{reg}_{\mathrm{DB}} : H^1_{\mathcal{M}}(X,\mathbb{Z}(1)) \to H^1_{\mathrm{DB}}(X,\mathbb{R}(1))$ sends f to $\log|f|$.

For example, take $X = \mathbb{G}_{m/\mathbb{C}}$ with coordinate t. Then $H^1_{\mathcal{M}}(X,\mathbb{Z}(1)) = \mathbb{C}^\times \cdot t^{\mathbb{Z}}$ and the regulator of the function $f = \lambda t^b$ is $\log|f| = \log|\lambda| + b\log|t|$. On the other hand, we saw in Exercise A.7 that the group $H^1_{\mathrm{DB}}(X,\mathbb{R}(1))$ consists of those functions of the form $a + b\log|t|$ with $a, b \in \mathbb{R}$. So in this particular case, the image of the regulator map $\mathrm{reg}_{\mathrm{DB}}$ is co-compact in $H^1_{\mathrm{DB}}(X,\mathbb{R}(1))$.

References

[1] N. H. Abel, Note sur la fonction $\psi x = x + \frac{x^2}{2^2} + \frac{x^3}{3^2} + \cdots + \frac{x^n}{n^2} + \cdots$, in *Œuvres complètes*, Vol. 2 (Imprimerie de Grondahl & Son, Christiania, 1881), 189–193.

[2] V. S. Adamchik, Integral and series representations for Catalan's constant, *Note*, www.cs.cmu.edu/~adamchik/articles/catalan.htm.

[3] S. Ahlgren, B. C. Berndt, A. J. Yee and A. Zaharescu, Integrals of Eisenstein series and derivatives of *L*-functions, *Intern. Math. Res. Not.* **2002** (2002), 1723–1738.

[4] H. Akatsuka, Zeta Mahler measures, *J. Number Theory* **129** (2009), 2713–2734.

[5] F. Amoroso and R. Dvornicich, A lower bound for the height in abelian extensions, *J. Number Theory* **80** (2000), 260–272.

[6] Y. André, *Une introduction aux motifs (motifs purs, motifs mixtes, périodes)*, Panoramas et Synthèses **17** (Société Mathématique de France, Paris, 2004).

[7] J. Ayoub, A guide to (étale) motivic sheaves, in *Proc. Intern. Congress Math.* (Seoul 2014), Vol. II (Kyung Moon Sa, Seoul, 2014), 1101–1124.

[8] D. H. Bailey and J. M. Borwein, Hand-to-hand combat with thousand-digit integrals, *J. Comput. Sci.* **3** (2012), 77–86.

[9] W. N. Bailey, *Generalized hypergeometric series*, Cambridge Tracts in Math. 32 (Cambridge University Press, Cambridge, 1935).

[10] H. F. Baker, Examples of the application of Newton's polygon to the theory of singular points of algebraic functions, *Trans. Cambridge Philos. Soc.* **15** (1893), 403–450.

[11] P. Balmer and M. Schlichting, Idempotent completion of triangulated categories, *J. Algebra* **236** (2001), no. 2, 819–834.

[12] T. Beck and J. Schicho, Curve parametrization over optimal field extensions exploiting the Newton polygon, in *Geometric modeling and algebraic geometry* (Springer, Berlin, 2008), 119–140.

[13] A. A. Beĭlinson, Higher regulators and values of *L*-functions of curves, *Funktsional. Anal. i Prilozhen.* **14** (1980), no. 2, 46–47; English transl., *Functional Anal. Appl.* **14** (1980), no. 2, 116–118.

[14] A. A. Beĭlinson, Higher regulators and values of *L*-functions, in *Current problems in mathematics* **24**, Itogi Nauki i Tekhniki (Akad. Nauk SSSR Vsesoyuz. Inst. Nauchn. i Tekhn.y Inform., Moscow, 1984), 181–238.

153

[15] A. A. Beĭlinson, Higher regulators of modular curves, in *Applications of algebraic K-theory to algebraic geometry and number theory* (Boulder, CO, 1983), Contemp. Math. **55** (Amer. Math. Soc., Providence, RI, 1986), 1–34.

[16] A. A. Beĭlinson, Notes on absolute Hodge cohomology, in *Applications of algebraic K-theory to algebraic geometry and number theory* (Boulder, CO, 1983), Contemp. Math. **55** (Amer. Math. Soc., Providence, RI, 1986), 35–68.

[17] B. C. Berndt, *Ramanujan's notebooks. Part IV* (Springer, New York, 1994).

[18] B. C. Berndt and A. Zaharescu, An integral of Dedekind eta-functions in Ramanujan's lost notebook, *J. Reine Angew. Math.* **551** (2002), 33–39.

[19] M. J. Bertin, Mesure de Mahler d'une famille de polynômes, *J. Reine Angew. Math.* **569** (2004), 175–188.

[20] M. J. Bertin, Mesure de Mahler d'hypersurfaces $K3$, *J. Number Theory* **128** (2008), 2890–2913.

[21] M. J. Bertin and O. Lecacheux, Elliptic fibrations on the modular surface associated to $\Gamma_1(8)$, in *Arithmetic and geometry of K3 surfaces and Calabi–Yau threefolds*, Fields Inst. Commun. **67** (Springer, New York, 2013), 153–199.

[22] M. J. Bertin and W. Zudilin, On the Mahler measure of a family of genus 2 curves, *Math. Z.* **283** (2016), 1185–1193.

[23] M. J. Bertin and W. Zudilin, On the Mahler measure of hyperelliptic families, *Ann. Math. Qué.* **41** (2017), 199–211.

[24] A. Besser and C. Deninger, p-adic Mahler measures, *J. Reine Angew. Math.* **517** (1999), 19–50.

[25] S. Bloch, The dilogarithm and extensions of Lie algebras, in *Algebraic K-theory* (Proc. Conf., Northwestern Univ., Evanston, Ill., 1980), Lecture Notes in Math. **854** (Springer, Berlin–New York, 1981), 1–23.

[26] S. Bloch, Algebraic cycles and the Beĭlinson conjectures, in *The Lefschetz centennial conference*, Part I (Mexico City, 1984), Contemp. Math. **58** (Amer. Math. Soc., Providence, RI, 1986), 65–79.

[27] S. J. Bloch, Higher regulators, algebraic K-theory, and zeta functions of elliptic curves, *Lecture notes* (UC Irvine, 1977); CRM Monograph Ser. **11** (Amer. Math. Soc., Providence, RI, 2000).

[28] S. Bloch and K. Kato, L-functions and Tamagawa numbers of motives, in *The Grothendieck Festschrift*, Vol. I, Progr. Math. **86** (Birkhäuser, Boston, MA, 1990), 333–400.

[29] S. Bloch and A. Ogus, Gersten's conjecture and the homology of schemes, *Ann. Sci. École Norm. Sup. (4)* **7** (1974), 181–201.

[30] L. A. Borisov and P. E. Gunnells, Toric modular forms of higher weight, *J. Reine Angew. Math.* **560** (2003), 43–64.

[31] H. Bornhorn, Mahler measures, K-theory and values of L-functions, *Preprint* arXiv:1503.06069 [math.NT].

[32] D. Borwein, J. M. Borwein, A. Straub and J. Wan, Log-sine evaluations of Mahler measures, II, *Integers* **12** (2012), 1179–1212.

[33] J. Borwein, A. van der Poorten, J. Shallit and W. Zudilin, *Neverending fractions. An introduction to continued fractions*, Austral. Math. Soc. Lecture Ser. **23** (Cambridge University Press, Cambridge, 2014).

[34] J. M. Borwein, A. Straub, J. Wan and W. Zudilin, Densities of short uniform random walks, with an appendix by D. Zagier, *Canad. J. Math.* **64** (2012), 961–990.

[35] P. Borwein, E. Dobrowolski and M. J. Mossinghoff, Lehmer's problem for polynomials with odd coefficients, *Ann. of Math. (2)* **166** (2007), 347–366.

[36] R. Bott and L. W. Tu, *Differential forms in algebraic topology*, Graduate Texts in Math. **82** (Springer, New York–Berlin, 1982).

[37] D. W. Boyd, Kronecker's theorem and Lehmer's problem for polynomials in several variables, *J. Number Theory* **13** (1981), 116–121.

[38] D. W. Boyd, Speculations concerning the range of Mahler's measure, *Canad. Math. Bull.* **24** (1981), 453–469.

[39] D. W. Boyd, Mahler's measure and special values of *L*-functions, *Experiment. Math.* **7** (1998), 37–82.

[40] D. W. Boyd, Uniform approximation to Mahler's measure in several variables, *Canad. Math. Bull.* **41** (1998), 125–128.

[41] D. W. Boyd, Mahler's measure and *L*-functions of elliptic curves evaluated at $s = 3$, *Slides from a lecture at the SFU/UBC number theory seminar* (December 7, 2006), www.math.ubc.ca/~boyd/sfu06.ed.pdf.

[42] D. Boyd, D. Lind, F. Rodriguez Villegas and C. Deninger, The many aspects of Mahler's measure, Final report of the Banff workshop 03w5035 (26 April–1 May 2003), www.birs.ca/workshops/2003/03w5035/report03w5035.pdf.

[43] D. Boyd and F. Rodriguez Villegas, Mahler's measure and the dilogarithm. I, *Canad. J. Math.* **54** (2002), 468–492.

[44] M. Branković, *Lehmer's problem and the Newton polygon*, BSc thesis (University of Newcastle, NSW, Australia, 2016).

[45] D. M. Bressoud, An easy proof of the Rogers–Ramanujan identities, *J. Number Theory* **16** (1983), 235–241.

[46] R. Breusch, On the distribution of the roots of a polynomial with integral coefficients, *Proc. Amer. Math. Soc.* **2** (1951), 939–941.

[47] D. Broadhurst, Feynman integrals, *L*-series and Kloosterman moments, *Comm. Number Theory Phys.* **10** (2016), 527–569.

[48] F. Brunault, Version explicite du théorème de Beilinson pour la courbe modulaire $X_1(N)$, *C. R. Math. Acad. Sci. Paris* **343** (2006), 505–510.

[49] F. Brunault, Beilinson–Kato elements in K_2 of modular curves, *Acta Arith.* **134** (2008), 283–298.

[50] F. Brunault, Parametrizing elliptic curves by modular units, *J. Aust. Math. Soc.* **100** (2016), 33–41.

[51] F. Brunault, Regulators of Siegel units and applications, *J. Number Theory* **163** (2016), 542–569.

[52] F. Brunault, Régulateurs modulaires explicites via la méthode de Rogers–Zudilin, *Compos. Math.* **153** (2017), 1119–1152.

[53] F. Brunault and M. Neururer, Mahler measures of elliptic modular surfaces, *Trans. Amer. Math. Soc.* **372** (2019), 119–152.

[54] J. I. Burgos, A C^∞ logarithmic Dolbeault complex, *Compositio Math.* **92** (1994), no. 1, 61–86.

[55] J. I. Burgos, Arithmetic Chow rings and Deligne–Beilinson cohomology, *J. Algebraic Geom.* **6** (1997), no. 2, 335–377.

[56] D. C. Cantor and E. G. Straus, On a conjecture of D. H. Lehmer, *Acta Arith.* **42** (1982), 97–100; Correction, *Acta Arith.* **42** (1983), 327.

[57] C. Carathéodory, Untersuchungen über die konformen Abbildungen von festen und veränderlichen Gebieten, *Math. Ann.* **72** (1912), 107–144.

[58] D.-C. Cisinski and F. Déglise, Mixed Weil cohomologies, *Adv. Math.* **230** (2012), no. 1, 55–130.

[59] D.-C. Cisinski and F. Déglise, *Triangulated categories of motives*, Springer Monographs in Math. (Springer, Cham, Switzerland, 2019).

[60] Th. Clausen, Beitrag zur Theorie der Reihen, *J. Reine Angew. Math.* **3** (1828), 92–95.

[61] J. D. Condon, *Mahler measure evaluations in terms of polylogarithms*, Dissertation (University of Texas at Austin, 2004).

[62] K. Conrad, Roots on a circle, *Expository note*, https://kconrad.math.uconn.edu/blurbs/galoistheory/numbersoncircle.pdf.

[63] F. Déglise and N. Mazzari, The rigid syntomic ring spectrum, *J. Inst. Math. Jussieu* **14** (2015), no. 4, 753–799.

[64] P. Deligne, Formes modulaires et représentations l-adiques, in *Séminaire Bourbaki*, Exposé no. 355, Lecture Notes in Math. **175** (Springer, Berlin, 1971), 139–172.

[65] P. Deligne, Théorie de Hodge. II, *Inst. Hautes Études Sci. Publ. Math.* **40** (1971), 5–57.

[66] P. Deligne, La conjecture de Weil. I, *Inst. Hautes Études Sci. Publ. Math.* **43** (1974), 273–307.

[67] P. Deligne, Valeurs de fonctions L et périodes d'intégrales, with an appendix by N. Koblitz and A. Ogus, in *Automorphic forms, representations and L-functions*, Proc. Sympos. Pure Math. **33**.2 (Amer. Math. Soc., Providence, RI, 1979), 313–346.

[68] C. Deninger, Deligne periods of mixed motives, K-theory and the entropy of certain $\mathbb{Z}^n$-actions, *J. Amer. Math. Soc.* **10** (1997), 259–281.

[69] C. Deninger, Extensions of motives associated to symmetric powers of elliptic curves and to Hecke characters of imaginary quadratic fields, in *Arithmetic geometry* (Cortona, 1994), Sympos. Math. **37** (Cambridge University Press, Cambridge, 1997), 99–137.

[70] C. Deninger and A. J. Scholl, The Beĭlinson conjectures, in *L-functions and arithmetic* (Durham, 1989), London Math. Soc. Lecture Note Ser. **153** (Cambridge University Press, Cambridge, 1991), 173–209.

[71] V. Dimitrov, A proof of the Schinzel–Zassenhaus conjecture on polynomials, *Preprint* arXiv:1912.12545 [math.NT].

[72] V. Dimitrov and P. Habegger, Galois orbits of torsion points near atoral sets, *Preprint* arXiv:1909.06051 [math.NT].

[73] J. D. Dixon, How good is Hadamard's inequality for determinants?, *Canad. Math. Bull.* **27** (1984), 260–264.

[74] J. D. Dixon and A. Dubickas, The values of Mahler measures, *Mathematika* **51** (2004), 131–148.

[75] E. Dobrowolski, On a question of Lehmer and the number of irreducible factors of a polynomial, *Acta Arith.* **34** (1979), 391–401.

[76] T. Dokchitser, Models of curves over DVRs, *Preprint* arXiv:1807.00025 [math.NT].

[77] C. F. Doran and M. Kerr, Algebraic K-theory of toric hypersurfaces, *Commun. Number Theory Phys.* **5** (2011), 397–600.

[78] V. G. Drinfeld, Two theorems on modular curves. *Funkcional. Anal. i Priložen.* **7** (1973), no. 2, 83–84.

[79] A. Dubickas, On the discriminant of the power of an algebraic number, *Studia Sci. Math. Hungar.* **44** (2007), 27–34.

[80] A. Dubickas and M. J. Mossinghoff, Auxiliary polynomials for some problems regarding Mahler's measure. *Acta Arith.* **119** (2005), 65–79.

[81] W. Duke and Ö. Imamoḡlu, On a formula of Bloch, *Funct. Approx.* **37** (2007), part 1, 109–117.

[82] H. Esnault and E. Viehweg, Deligne–Beĭlinson cohomology, in *Beĭlinson's conjectures on special values of L-functions*, Perspect. Math. **4** (Academic Press, Boston, MA, 1988), 43–91.

[83] G. Everest and T. Ward, *Heights of polynomials and entropy in algebraic dynamics*, Universitext (Springer, London, 1999).

[84] L. D. Faddeev and R. M. Kashaev, Quantum dilogarithm, *Modern Phys. Lett. A* **9** (1994), 427–434.

[85] Ch. Fan and F. Y. Wu, General lattice model of phase transitions *Phys. Rev. B* **2** (1970), 723–733.

[86] M. Flach, The equivariant Tamagawa number conjecture: a survey, with an appendix by C. Greither, in *Stark's conjectures: recent work and new directions*, Contemp. Math. **358** (Amer. Math. Soc., Providence, RI, 2004), 79–125.

[87] J.-M. Fontaine and B. Perrin-Riou, Autour des conjectures de Bloch et Kato: cohomologie galoisienne et valeurs de fonctions L, in *Motives* (Seattle, WA, 1991), Proc. Sympos. Pure Math. **55** (Amer. Math. Soc., Providence, RI, 1994), 599–706.

[88] W. Fulton, *Introduction to toric varieties*, Annals of Math. Studies **131**, The William H. Roever Lectures in Geometry (Princeton University Press, Princeton, NJ, 1993).

[89] G. Gasper and M. Rahman, *Basic hypergeometric series*, 2nd edn., Encyclopedia Math. Appl. **96** (Cambridge University Press, Cambridge, 2004).

[90] C. F. Gauss, Articles 356 and 357 in *Untersuchungen über höhere Arithmetik* (Chelsea Publishing, New York, 1965).

[91] M. T. Gealy, *On the Tamagawa number conjecture for motives attached to modular forms*, PhD thesis (California Institute of Technology, 2005), https://resolver.caltech.edu/CaltechETD:etd-12162005-124435.

[92] S. I. Gelfand and Y. I. Manin, *Methods of homological algebra*, 2nd edn., Springer Monographs in Math. (Springer, Berlin, 2003).

[93] E. Ghate and E. Hironaka, The arithmetic and geometry of Salem numbers, *Bull. Amer. Math. Soc.* **38** (2001), 293–314.

[94] S. P. Glasby, Cyclotomic ordering conjecture, *Preprint* arXiv:1903.02951 [math.NT].

[95] M. Griffin, K. Ono, L. Rolen and D. Zagier, Jensen polynomials for the Riemann zeta function and other sequences, *Proc. Nat. Acad. Sci. USA* **116** (2019), no. 23, 11103–11110.

[96] A. GROTHENDIECK, On the de Rham cohomology of algebraic varieties, *Inst. Hautes Études Sci. Publ. Math.* **29** (1966), 95–103.

[97] J. GUILLERA and W. ZUDILIN, Ramanujan-type formulae for $1/\pi$: the art of transla-tion, in *The Legacy of Srinivasa Ramanujan*, B. C. Berndt and D. Prasad (eds.), Ramanujan Math. Soc. Lecture Notes Ser. **20** (2013), 181–195.

[98] A. GUILLOUX and J. MARCHÉ, Volume function and Mahler measure of exact poly-nomials, *Preprint* arXiv:1804.01395 [math.GT].

[99] A. HATCHER, *Algebraic topology* (Cambridge University Press, Cambridge, 2002).

[100] G. HÖHN, On a theorem of Garza regarding algebraic numbers with real conju-gates, *Intern. J. Number Theory* **7** (2011), 943–945.

[101] G. HÖHN and N.-P. SKORUPPA, Un résultat de Schinzel, *J. Théor. Nombres Bor-deaux* **5** (1993), 185.

[102] A. HOLMSTROM and J. SCHOLBACH, Arakelov motivic cohomology I, *J. Algebraic Geom.* **24** (2015), no. 4, 719–754.

[103] A. HUBER and G. KINGS, Dirichlet motives via modular curves, *Ann. Sci. École Norm. Sup. (4)* **32** (1999), no. 3, 313–345.

[104] D. HUYBRECHTS, *Complex geometry. An introduction*, Universitext (Springer, Berlin, 2005).

[105] U. JANNSEN, Deligne homology, Hodge-$\mathcal{D}$-conjecture, and motives, in *Beĭlinson's conjectures on special values of L-functions*, Perspect. Math. **4** (Academic Press, Boston, MA, 1988), 305–372.

[106] U. JANNSEN, *Mixed motives and algebraic K-theory*, Lecture Notes in Math. **1400** (Springer, Berlin, 1990).

[107] J. L. W. V. JENSEN, Sur un nouvel et important théorème de la théorie des fonc-tions, *Acta Math.* **22** (1899), 359–364.

[108] C.-G. JI and W.-P. LI, Values of coefficients of cyclotomic polynomials, *Discr. Math.* **308** (2008), 5860–5863.

[109] B. KAHN, *Fonctions zêta et L de variétés et de motifs*, Nano (Calvage et Mounet, Paris, 2018).

[110] R. M. KASHAEV, The q-binomial formula and the Rogers dilogarithm iden-tity, *Vestn. Chelyab. Gos. Univ. Mat. Mekh. Inform.* (2015), no. 3 (17), 62–66; *Preprint* arXiv:math.QA/0407078.

[111] K. KATO, p-adic Hodge theory and values of zeta functions of modular forms, *Astérisque* **295** (2004), 117–290.

[112] K. KHURI-MAKDISI, Moduli interpretation of Eisenstein series, *Int. J. Number Theory* **8** (2012), 715–748.

[113] M. KRAITCHIK, *Recherches sur la théorie des nombres*, tome I (Gauthier-Villars, Paris, 1924); tome II (Gauthier-Villars, Paris, 1929).

[114] C. KRATTENTHALER, Advanced determinant calculus, *Sém. Lothar. Combin.* **42** (1999), 67 pp.

[115] N. KUROKAWA, M. N. LALÍN and H. OCHIAI, Higher Mahler measures and zeta functions, *Acta Arith.* **135** (2008), 269–297.

[116] N. KUROKAWA and H. OCHIAI, Mahler measures via the crystalization, *Comment. Math. Univ. St. Pauli* **54** (2005), 121–137.

[117] M. N. LALÍN, An algebraic integration for Mahler measure, *Duke Math. J.* **138** (2007), 391–422.

[118] M. N. Lalín, Mahler measure and elliptic curve *L*-functions at $s = 3$, *J. Reine Angew. Math.* **709** (2015), 201–218.

[119] M. N. Lalín and M. D. Rogers, Functional equations for Mahler measures of genus-one curves, *Algebra and Number Theory* **1** (2007), 87–117.

[120] W. M. Lawton, A generalization of a theorem of Kronecker, *J. Sci. Faculty Chiangmai Univ. (Thailand)* **4** (1977), 15–23.

[121] W. M. Lawton, A problem of Boyd concerning geometric means of polynomials, *J. Number Theory* **16** (1983), 356–362.

[122] D. H. Lehmer, Factorization of certain cyclotomic functions, *Ann. of Math.* **34** (1933), 461–479.

[123] D. Lind, Lehmer's problem for compact abelian groups, *Proc. Amer. Math. Soc.* **133** (2005), 1411–1416.

[124] D. Lind, K. Schmidt and E. Verbitskiy, Homoclinic points, atoral polynomials, and periodic points of algebraic $\mathbb{Z}^d$-actions, *Ergodic Theory Dynam. Systems* **33** (2013), 1060–1081.

[125] H. Liu, Data for defining tempered genus 1 families and *j*-invariant of these families (2019), https://github.com/liuhangsnnu/mahler-measure-of-genus-2-and-3-curves.

[126] The LMFDB Collaboration, The *L*-functions and modular forms database, www.lmfdb.org (2013–2020).

[127] R. Louboutin, Sur la mesure de Mahler d'un nombre algébrique, *C. R. Acad. Sci. Paris Sér. I Math.* **296** (1983), 707–708.

[128] J. H. Loxton, Special values of the dilogarithm function, *Acta Arith.* **43** (1984), 155–166.

[129] K. Mahler, An application of Jensen's formula to polynomials, *Mathematika* **7** (1960), 98–100.

[130] K. Mahler, On some inequalities for polynomials in several variables, *J. London Math. Soc.* **37** (1962), 341–344.

[131] K. Mahler, An inequality for the discriminant of a polynomial, *Michigan Math. J.* **11** (1964), 257–262.

[132] V. Maillot, Géométrie d'Arakelov des variétés toriques et fibrés en droites intégrables, *Mém. Soc. Math. Fr. (N.S.)* **80** (2000).

[133] Y. I. Manin, Parabolic points and zeta functions of modular curves, *Math. USSR Izvestija* **6** (1972), no. 1, 19–64.

[134] J. Martin, *Building infinite ray-class towers with specific signatures and small bounded root discriminant*, PhD dissertation (Cornell University, 2006).

[135] E. M. Matveev, A relationship between the Mahler measure and the discriminant of algebraic numbers, *Mat. Zametki* **59** (1996), 415–420; English transl., *Math. Notes* **59** (1996), 293–297.

[136] C. Mazza, V. Voevodsky and C. Weibel, *Lecture notes on motivic cohomology*, Clay Math. Monographs **2** (Amer. Math. Soc., Providence, RI; Clay Math. Institute, Cambridge, MA, 2006).

[137] A. Mellit, Elliptic dilogarithms and parallel lines, *J. Number Theory* **204** (2019), 1–24.

[138] *Great Internet Mersenne Prime Search*, www.mersenne.org.

[139] M. Mignotte, An inequality about factors of polynomials, *Math. Comp.* **28** (1974), 1153–1157.

[140] J. MILNOR, *Introduction to algebraic K-theory*, Ann. of Math. Studies **72** (Princeton University Press, Princeton, NJ, 1971).

[141] H. MINKOWSKI, Über die positiven quadratischen Formen und über kettenbruchähnliche Algorithmen, *J. Reine Angew. Math.* **107** (1891), 278–297.

[142] T. MIYAKE, *Modular forms* (Springer, Berlin, 1989).

[143] F. MOREL and V. VOEVODSKY, $\mathbb{A}^1$-homotopy theory of schemes, *Inst. Hautes Études Sci. Publ. Math.* **90** (1999), 45–143.

[144] M. J. MOSSINGHOFF, *Lehmer's problem web page*, www.cecm.sfu.ca/~mjm/Lehmer/lc.html.

[145] J. P. MURRE, J. NAGEL and C. A. M. PETERS, *Lectures on the theory of pure motives*, Univ. Lecture Ser. **61** (Amer. Math. Soc., Providence, RI, 2013).

[146] J. NEKOVÁŘ, Beĭlinson's conjectures, in *Motives* (Seattle, WA, 1991), Proc. Sympos. Pure Math. **55** (Amer. Math. Soc., Providence, RI, 1994), 537–570.

[147] Y. V. NESTERENKO, Integral identities and constructions of approximations to zeta-values, *J. Théor. Nombres Bordeaux* **15** (2003), 535–550.

[148] A. M. ODLYZKO, Bounds for discriminants and related estimates for class numbers, regulators and zeros of zeta functions: a survey of recent results, *Sém. Théor. Nombres Bordeaux (2)* **2** (1990), 119–141.

[149] L. ONSAGER, Crystal statistics. I. A two-dimensional model with an order-disorder transition, *Phys. Rev.* **65** (1944), 117–149.

[150] M. A. PAPANIKOLAS, M. D. ROGERS and D. SAMART, The Mahler measure of a Calabi–Yau threefold and special L-values, *Math. Z.* **276** (2014), 1151–1163.

[151] PARI/GP, https://pari.math.u-bordeaux.fr (1990–2020).

[152] T. A. PIERCE, Numerical factors of the arithmetic forms $\prod_{i=1}^{n}(1 \pm \alpha_i^m)$, *Ann. of Math.* **18** (1917), 53–64.

[153] C. PINNER and C. SMYTH, Integer group determinants for small groups, *Ramanujan J.* **51** (2020), no. 2, 421–453.

[154] G. PÓLYA and G. SZEGŐ, *Problems and theorems in analysis. I. Series, integral calculus, theory of functions*, Grundlehren Math. Wiss. **193** (Springer, Berlin–New York, 1978).

[155] G. PÓLYA and G. SZEGŐ, *Problems and theorems in analysis. II. Theory of functions, zeros, polynomials, determinants, number theory, geometry*, Grundlehren Math. Wiss. **216** (Springer, New York–Heidelberg, 1976).

[156] C. POMERANCE and S. RUBINSTEIN-SALZEDO, Cyclotomic coincidences, *Exp. Math.* (to appear); doi:10.1080/10586458.2019.1660741.

[157] V. V. PRASOLOV, *Polynomials*, Algorithms Comput. Math. **11** (Springer, Berlin, 2004).

[158] D. QUILLEN, Higher algebraic K-theory. I, in *Algebraic K-theory, I: Higher K-theories* (Proc. Conf., Battelle Memorial Inst., Seattle, WA, 1972), Lecture Notes in Math. **341** (Springer, Berlin, 1973), 85–147.

[159] S. RABINOWITZ, A census of convex lattice polygons with at most one interior lattice point, *Ars Combin.* **28** (1989), 83–96.

[160] D. RAMAKRISHNAN, A regulator for curves via the Heisenberg group, *Bull. Amer. Math. Soc. (N.S.)* **5** (1981), no. 6, 191–195.

[161] D. RAMAKRISHNAN, Regulators, algebraic cycles, and values of L-functions, in *Algebraic K-theory and algebraic number theory* (Honolulu, HI, 1987), Contemp. Math. **83** (Amer. Math. Soc., Providence, RI, 1989), 183–310.

[162] B. RICHMOND and G. SZEKERES, Some formulas related to dilogarithms, the zeta function and the Andrews–Gordon identities, *J. Austral. Math. Soc. Ser. A* **31** (1981), 362–373.

[163] H. RIESEL, *Prime numbers and computer methods for factorization*, 2nd edn., Progress in Math. **126** (Birkhäuser, Boston, MA, 1994).

[164] F. RODRIGUEZ VILLEGAS, Modular Mahler measures, *Preprint* (November 1996), 10 pp.

[165] F. RODRIGUEZ VILLEGAS, Modular Mahler measures. I, in *Topics in number theory* (University Park, PA, 1997), Math. Appl. **467** (Kluwer Acad. Publ., Dordrecht, 1999), 17–48.

[166] L. J. ROGERS, On function sum theorems connected with the series $\sum_{n=1}^{\infty} x^n/n^2$, *Proc. London Math. Soc.* **4** (1907), 169–189.

[167] M. ROGERS, New $_5F_4$ hypergeometric transformations, three-variable Mahler measures, and formulas for $1/\pi$, *Ramanujan J.* **18** (2009), 327–340.

[168] M. ROGERS, Hypergeometric formulas for lattice sums and Mahler measures, *Intern. Math. Res. Not.* **2011** (2011), 4027–4058.

[169] M. ROGERS and W. ZUDILIN, From L-series of elliptic curves to Mahler measures, *Compositio Math.* **148** (2012), 385–414.

[170] M. ROGERS and W. ZUDILIN, On the Mahler measure of $1 + X + 1/X + Y + 1/Y$, *Intern. Math. Res. Not.* **2014** (2014), 2305–2326.

[171] R. SALEM, A remarkable class of algebraic integers. Proof of a conjecture of Vijayaraghavan, *Duke Math. J.* **11** (1944), 103–108.

[172] D. SAMART, Three-variable Mahler measures and special values of modular and Dirichlet L-series, *Ramanujan J.* **32** (2013), 245–268.

[173] D. SAMART, The elliptic trilogarithm and Mahler measures of $K3$ surfaces, *Forum Math.* **28** (2016), 405–423.

[174] N. SCHAPPACHER and A. J. SCHOLL, Beĭlinson's theorem on modular curves, in *Beĭlinson's conjectures on special values of L-functions*, Perspect. Math. **4** (Academic Press, Boston, MA, 1988), 273–304.

[175] A. SCHINZEL, On the product of the conjugates outside the unit circle of an algebraic number, *Acta Arith.* **24** (1973), 385–399; Addendum, *Acta Arith.* **26** (1974/75), 329–331.

[176] A. SCHINZEL, *Polynomials with special regard to reducibility*, with an appendix by U. Zannier, Encyclopedia Math. Appl. **77** (Cambridge University Press, Cambridge, 2000).

[177] A. SCHINZEL and H. ZASSENHAUS, A refinement of two theorems of Kronecker, *Michigan Math. J.* **12** (1965), 81–85.

[178] K. SCHMIDT and E. VERBITSKIY, New directions in algebraic dynamical systems, *Regul. Chaotic Dyn.* **16** (2011), 79–89.

[179] P. SCHNEIDER, Introduction to the Beĭlinson conjectures, in *Beĭlinson's conjectures on special values of L-functions*, Perspect. Math. **4** (Academic Press, Boston, MA, 1988), 1–35.

[180] A. J. SCHOLL, Classical motives, in *Motives* (Seattle, WA, 1991), Proc. Sympos. Pure Math. **55** (Amer. Math. Soc., Providence, RI, 1994), 163–187.

[181] A. J. SCHOLL, Integral elements in K-theory and products of modular curves, in *The arithmetic and geometry of algebraic cycles* (Banff, AB, 1998), NATO Sci. Ser. C Math. Phys. Sci. **548** (Kluwer Acad. Publ., Dordrecht, 2000), 467–489.

[182] A. Scholz, Minimaldiskriminanten algebraischer Zahlkörper, *J. Reine Angew. Math.* **179** (1938), 16–21.

[183] M.-P. Schützenberger, Une interprétation de certaines solutions de l'équation fonctionnelle: $F(x+y) = F(x)F(y)$, *C. R. Acad. Sci. Paris* **236** (1953), 352–353.

[184] P. R. Scott, On convex lattice polygons, *Bull. Austral. Math. Soc.* **15** (1976), 395–399.

[185] J.-P. Serre, *Cours d'arithmétique*, Collection SUP: "Le Mathématicien" 2 (Presses Universitaires de France, Paris, 1970).

[186] H. N. Shapiro and G. H. Sparer, On the units of cubic fields, *Comm. Pure Appl. Math.* **26** (1973), 819–835.

[187] G. Shimura, *Introduction to the arithmetic theory of automorphic functions*, Kan Memorial Lectures. Publications of the Mathematical Society of Japan **11** (Iwanami Shoten, Publishers, Tokyo and Princeton University Press, Princeton, 1971).

[188] E. Shinder and M. Vlasenko, Linear Mahler measures and double *L*-values of modular forms, *J. Number Theory* **142** (2014), 149–182.

[189] C. L. Siegel, Algebraic integers whose conjugates lie in the unit circle, *Duke Math. J.* **11** (1944), 597–602.

[190] C. L. Siegel, Some remarks on discontinuous groups, *Ann. of Math.* **46** (1945), 708–718.

[191] C. L. Siegel, *Advanced analytic number theory*, Tata Inst. Lecture Notes Ser. (Tata Institute of Fundamental Research, India, 1961), www.math.tifr.res.in/~publ/ln/tifr23.pdf .

[192] J. H. Silverman, *Advanced topics in the arithmetic of elliptic curves*, Graduate Texts in Mathematics **151** (Springer, New York, 1994).

[193] D. Simon, Construction de polynômes de petits discriminants, *C. R. Acad. Sci. Paris Sér. I Math.* **329** (1999), no. 6, 465–468.

[194] L. J. Slater, *Generalized hypergeometric functions* (Cambridge University Press, Cambridge, 1966).

[195] C. J. Smyth, On the product of the conjugates outside the unit circle of an algebraic integer, *Bull. London Math. Soc.* **3** (1971), 169–175.

[196] C. J. Smyth, *Topics in the theory of numbers*, PhD thesis (Cambridge University, Cambridge, 1972).

[197] C. J. Smyth, A Kronecker-type theorem for complex polynomials in several variables, *Canad. Math. Bull.* **24** (1981), 447–452.

[198] C. J. Smyth, The Mahler measure of algebraic numbers: a survey, in *Number theory and polynomials*, London Math. Soc. Lecture Note Ser. **352** (Cambridge University Press, Cambridge, 2008), 322–349.

[199] C. J. Smyth, Seventy years of Salem numbers, *Bull. London Math. Soc.* **47** (2015), 379–395.

[200] W. Spence, *Essay on the theory of the various orders of logarithmic transcendents: with an inquiry into their applications to the integral calculus and the summation of series* (Murray/Constable, London–Edinburgh, 1809).

[201] W. Stein, *Modular forms, a computational approach*, with an appendix by P. E. Gunnells, Graduate Studies in Mathematics **79** (Amer. Math. Soc., Providence, RI, 2007), https://wstein.org/books/modform/modform/index.html .

[202] G. Stevens, The Eisenstein measure and real quadratic fields, in *Théorie des nombres*, Proc. Conf. (Quebec, PQ, 1987) (de Gruyter, Berlin, 1989), 887–927.

[203] A. Straub and W. Zudilin, Short walk adventures, in *From analysis to visualization: A celebration of the life and legacy of Jonathan M. Borwein* (Callaghan, Australia, September 2017), B. Sims et al. (eds.), Springer Proceedings in Math. Stat. **313** (Springer, New York, 2020), 423–439.

[204] J. Sturm, On the congruence of modular forms, in *Number theory* (New York, 1984–1985), Lecture Notes in Math. **1240** (Springer, Berlin, 1987), 275–280.

[205] J. Suzuki, On coefficients of cyclotomic polynomials *Proc. Japan Acad. Ser. A Math. Sci.* **63** (1987), no. 7, 279–280.

[206] G. Szegő, Ein Grenzwertsatz über die Toeplitzschen Determinanten einer reellen positiven Funktion, *Math. Ann.* **76** (1915), 490–503.

[207] R. Takloo-Bighash, A remark on a paper of S. Ahlgren, B. C. Berndt, A. J. Yee, and A. Zaharescu: "Integrals of Eisenstein series and derivatives of *L*-functions" [3], *Intern. J. Number Theory* **2** (2006), 111–114.

[208] W. Thurston, *Geometry and topology of three-manifolds*, Lecture notes (Princeton University, 1980), http://library.msri.org/books/gt3m/.

[209] N. Touafek, Mahler's measure: proof of two conjectured formulae, *An. Ştiinţ. Univ. "Ovidius" Constanţa Ser. Mat.* **16** (2008), 127–136.

[210] S. Vandervelde, A formula for the Mahler measure of $axy+bx+cy+d$, *J. Number Theory* **100** (2003), 184–202.

[211] E. Verbitskiy, Spanning trees, abelian sandpiles, and algebraic dynamical systems, in *Workshop on "Probabilistic models with determinantal structure"*, T. Shirai (ed.), Math-for-Industry Lecture Note Ser. **63** (Kyushu University, Fukuoka, 2015), 1–16.

[212] V. Voevodsky, Motivic cohomology groups are isomorphic to higher Chow groups in any characteristic, *Intern. Math. Res. Not.* **7** (2002), 351–355.

[213] V. Voevodsky, A. Suslin and E. M. Friedlander, *Cycles, transfers, and motivic homology theories*, Annals of Mathematics Studies **143** (Princeton University Press, Princeton, NJ, 2000).

[214] J. Voight, Enumeration of totally real number fields of bounded root discriminant, in *Algorithmic number theory*, Lecture Notes in Comput. Sci. **5011** (Springer, Berlin, 2008), 268–281.

[215] P. Voutier, An effective lower bound for the height of algebraic numbers, *Acta Arith.* **74** (1996), 81–95.

[216] S. O. Warnaar and W. Zudilin, Dedekind's η-function and Rogers–Ramanujan identities, *Bull. London Math. Soc.* **44** (2012), 1–11.

[217] M. Yoshida, *Hypergeometric functions, my love*, Modular interpretations of configuration spaces, Aspects Math. **E32** (Friedr. Vieweg & Sohn, Braunschweig, 1997).

[218] D. Zagier, Algebraic numbers close to both 0 and 1, *Math. Comp.* **61** (1993), no. 203, 485–491.

[219] D. Zagier, The dilogarithm function, in *Frontiers in number theory, physics, and geometry. II* (Springer, Berlin, 2007), 3–65.

[220] S. Zhang, Positive line bundles on arithmetic surfaces, *Ann. of Math. (2)* **136** (1992), 569–587.

[221] Y. ZHOU, Wick rotations, Eichler integrals, and multi-loop Feynman diagrams, *Comm. Number Theory Phys.* **12** (2018), 127–192.

[222] W. ZUDILIN, Period(d)ness of L-values, in *Number theory and related fields, in memory of Alf van der Poorten*, J. M. Borwein et al. (eds.), Springer Proceedings in Math. Stat. **43** (Springer, New York, 2013), 381–395.

[223] W. ZUDILIN, Regulator of modular units and Mahler measures, *Math. Proc. Cambridge Philos. Soc.* **156** (2014), 313–326.

Index